Lecture Notes in Computer Science 3247

Commenced Publication in 1973
Founding and Former Series Editors:
Gerhard Goos, Juris Hartmanis, and Jan van Leeuwen

W0193223

Dorin Comaniciu   Kenichi Kanatani
Rudolf Mester   David Suter (Eds.)

Statistical Methods in Video Processing

ECCV 2004 Workshop SMVP 2004
Prague, Czech Republic, May 16, 2004
Revised Selected Papers

 Springer

Volume Editors

Dorin Comaniciu
Siemens Corporate Research
Integrated Data Systems Department
755 College Road East, Princeton, NJ 08540, USA
E-mail: dorin.comaniciu@siemens.com

Kenichi Kanatani
Okayama University
Department of Information Technology
Okayama 700-8530, Japan
E-mail: kanatani@suri.it.okayama-u.ac.jp

Rudolf Mester
Johann-Wolfgang-Goethe Universität Frankfurt
Institute for Applied Physics
Robert-Mayer-Str. 2–4, 60054 Frankfurt/Main, Germany
E-mail: mester@iap.uni-frankfurt.de

David Suter
Monash University
Department of Electrical and Computer Systems Engineering
P.O. Box 35, Clayton 3800, VIC, Australia
E-mail: d.suter@eng.monash.edu.au

Library of Congress Control Number: 2004116560

CR Subject Classification (1998): I.4, I.3, I.5, G.3, F.2.1, I.2

ISSN 0302-9743
ISBN 3-540-23989-8 Springer Berlin Heidelberg New York

Springer is a part of Springer Science+Business Media

springeronline.com

© Springer-Verlag Berlin Heidelberg 2004
Printed in Germany

Typesetting: Camera-ready by author, data conversion by Scientific Publishing Services, Chennai, India
Printed on acid-free paper   SPIN: 11363644   06/3142   5 4 3 2 1 0

Preface

The 2nd International Workshop on Statistical Methods in Video Processing, SMVP 2004, was held in Prague, Czech Republic, as an associated workshop of ECCV 2004, the 8th European Conference on Computer Vision.

A total of 30 papers were submitted to the workshop. Of these, 17 papers were accepted for presentation and included in these proceedings, following a double-blind review process. The workshop had 42 registered participants.

The focus of the meeting was on recent progress in the application of advanced statistical methods to solve computer vision tasks. The one-day scientific program covered areas of high interest in vision research, such as dense reconstruction of 3D scenes, multibody motion segmentation, 3D shape inference, errors-in-variables estimation, probabilistic tracking, information fusion, optical flow computation, learning for nonstationary video data, novelty detection in dynamic backgrounds, background modeling, grouping using feature uncertainty, and crowd segmentation from video.

We wish to thank the authors of all submitted papers for their interest in the workshop. We also wish to thank the members of our program committee and the external reviewers for their commitment of time and effort in providing valuable recommendations for each submission. We are thankful to Vaclav Hlavac, the General Chair of ECCV 2004, and to Radim Sara, for the local organization of the workshop and registration management.

We hope you will find these proceedings both inspiring and of high scientific quality.

June 2004

Dorin Comaniciu
Kenichi Kanatani
Rudolf Mester
David Suter

Organization

Organizing Committee

Dorin Comaniciu   Siemens Corporate Research, USA
Kenichi Kanatani   Okayama University, Japan
Rudolf Mester   Goethe-Universitaet, Germany
David Suter   Monash University, Australia

Program Committee

Bir Bhanu   University of California, USA
Patrick Bouthemy   IRISA/INRIA, France
Mike Brooks   University of Adelaide, Australia
Yaron Caspi   Weizmann Institute of Science, Israel
Rama Chellappa   University of Maryland, USA
Andrew Fitzgibbon   Oxford University, UK
Radu Horaud   INRIA, France
Naoyuki Ichimura   AIST, Japan, and Columbia University, USA
Michael Isard   Microsoft Research, USA
Bogdan Matei   Sarnoff Corporation, USA
Takashi Matsuyama   Kyoto University, Japan
Visvanathan Ramesh   Siemens Corporate Research, USA
Harpreet Sawhney   Sarnoff Corporation, USA
Stuart Schwartz   Princeton University, USA
Mubarak Shah   University of Central Florida, USA
Nobutaka Shimada   Osaka University, Japan
Zhengyou Zhang   Microsoft Research, USA
Ying Zhu   Siemens Corporate Research, USA

External Reviewers

A. Bab-Hadiashar   P. Meer   A. Van Den Hengel
R. Koch   M. Muehlich

Sponsoring Institution

Siemens Corporate Research, Inc.
Princeton, NJ, USA

Table of Contents

Image/Video Analysis

Towards Complete Free-Form Reconstruction of Complex 3D Scenes from an Unordered Set of Uncalibrated Images

H. Cornelius[1], R. Šára[2], D. Martinec[2], T. Pajdla[2], O. Chum[2], and J. Matas[2]

[1] Royal Institute of Technology (KTH),
Department of Numerical Analysis and Computing Science,
100 44 Stockholm, Sweden
hugoc@nada.kth.se
http://www.nada.kth.se
[2] Center for Machine Perception, Czech Technical University,
166 27 Prague, Czech Republic
{sara, martid1, pajdla, chum, matas}@cmp.felk.cvut.cz
http://cmp.felk.cvut.cz

Abstract. This paper describes a method for accurate dense reconstruction of a complex scene from a small set of high-resolution unorganized still images taken by a hand-held digital camera. A fully automatic data processing pipeline is proposed. Highly discriminative features are first detected in all images. Correspondences are then found in all image pairs by wide-baseline stereo matching and used in a scene structure and camera reconstruction step that can cope with occlusion and outliers. Image pairs suitable for dense matching are automatically selected, rectified and used in dense binocular matching. The dense point cloud obtained as the union of all pairwise reconstructions is fused by local approximation using oriented geometric primitives. For texturing, every primitive is mapped on the image with the best resolution.

The global structure reconstruction in the first step allows us to work with an unorganized set of images and to avoid error accumulation. By using object-centered geometric primitives we are able to preserve the flexibility of the method to describe complex free-form structures, preserve the possibility to build the dense model in an incremental way, and to retain the possibility to refine the cameras and the dense model by bundle adjustment. Results are demonstrated on partial models of a circular church and a Henri de Miller's sculpture. We observed spatial resolution in the range of centimeters on objects of about 20 m in size.

1 Introduction

Building geometric representation of a complex scene from a set of views is one of the classical Computer Vision problems. The task is to obtain a model that (1) can either be used to generate a novel view for a moving observer or (2) contains explicit representation of the structure (3D topology and geometry) of the scene.

D. Comaniciu et al. (Eds.): SMVP 2004, LNCS 3247, pp. 1–12, 2004.

The focus of this paper is on the latter. We present a method that obtains the 3D model from a small unordered set of uncalibrated images. This means that the camera order, positions and (most of) their intrinsic parameters are not known. Hence, a method that is capable of *joint* estimation of the cameras and the scene structure must be used. This is usually done in two independent steps: a pre-calibration, which recovers the cameras from a sparse set of features, followed by scene structure estimation and densification. To increase the accuracy of both scene structure and the cameras, this can be followed by bundle adjustment.

In this paper we basically follow the cascade: (1) wide-baseline matching, (2) camera and 3D structure reconstruction, (3) dense matching, (4) 3D model reconstruction. By splitting the procedure to smaller, explicit blocks, we hope for achieving good performance by solving a well defined task at every step. Every such task can then use a different, optimal, prior model of the scene. The first stage uses discriminative image regions and a strong local planarity model to establish initial matches that are used in the second stage, where a consistent set of cameras are searched for under occlusion and outliers. Pinhole camera model with radial distortion is used. The subsequent matching then focuses on density and accuracy while minimizing false positives under missing texture and inaccurate epipolar geometry. The last step first approximates the point cloud density as a mixture of local kernels which helps maintain the efficiency of the subsequent processing. Since every stage makes only a small step in image interpretation, in our future work we will be able to close a feedback loop from almost every stage and refine the camera parameters and the scene structure before we commit to final interpretation of scene structure (like a triangulated model of certain topology).

Our system, described in detail in this paper, differs from other work by:

1. A new method for the reconstruction of the projective structure and the cameras using a *robust* and *global* optimization procedure that does not require the input images to form a sequence (a known-order image set). This means that reconstruction errors do not propagate.
2. Using an *object-centered* model of the *local* geometry of the scene, which gives the possibility (1) to process and model images of complex structures, (2) to grow the model in an incremental way while preserving the accuracy, (3) to insert high-resolution partial reconstructions into the model, (4) to perform an efficient iterative refinement (bundle adjustment) in the final accuracy-increasing step, (5) to impose local spatial coherence, and (6) to effectively *compress* the dense point cloud while preserving its ability to model complex free-form structures.

2   Method

In this section we describe the data processing pipeline. In brief the method works as follows. The input is a number of photographs of the object to be reconstructed. Sparse correspondences are searched for across all pairs of views. A consistent system of cameras is estimated and the cameras are auto-calibrated.

Image pairs suitable for dense matching are selected and rectified and used for dense matching. 3D points are reconstructed from each image pair and the union of the partial reconstructions form a dense point cloud. Local geometric models called fish-scales are fit to the point cloud and each fish-scale is texture-mapped from the image with the best resolution and view of that fish-scale.

Input. The input is a number of photographs of the object to be reconstructed, taken with a hand-held compact digital camera. For the method to work well, there should be image pairs taken with both wide and relatively narrow baseline among the images. The wide baseline pairs support the numerical stability of camera auto-calibration. If available, the narrow baseline pairs are more suitable for dense matching because stronger prior models (like ordering) can be used.

Our method makes it possible to use photographs of different resolution. To reconstruct the overall 3D structure of the object, overview images are used. Higher resolution images can be used to blend in parts of the object with fine geometric details or with texture too poor to suffice for reliable matching at the lower resolution. Examples of input data obtained under this scheme are shown in Figs. 1 and 2.

Fig. 1. Input images for the Head Scene

Region Matching. The first step is to find sparse correspondences across all images. This is done by matching maximally stable extremal regions (MSERs) [1] in all possible pairs of views. The epipolar geometry for each image pair is estimated using LO-RANSAC [2]. Taking only the matches satisfying the epipolar constraint, we get for every image pair a set of sparse correspondences with relatively few outliers with respect to the true scene structure. The ability of the method to handle large changes in scale and brightness is essential, since we are necessarily dealing with wide baseline photographs of varying resolution.

Suppose the object has two or more parts looking the same. To reduce the risk that too many matches between similar regions on different parts of the object would result in a wrong reconstruction of the cameras, it might seem necessary to forbid matching between images seeing these different parts. However, it is possible to phase-out most of such image pairs automatically, as described below.

An example of a set of detected MSERs in a wide baseline pair is shown in Fig. 3.

Fig. 2. Input images for the St. Martin scene. Note that there are narrow-baseline overview and close-up pairs that are mutually separated by wide baselines

Fig. 3. Maximally stable extremal regions (circles) detected in two wide-baseline pairs

Estimation of a Consistent System of Cameras. Assuming full perspective camera model, projection of each point $\mathbf{X}_p$ visible in camera $\mathbf{P}^i$ can be written in homogeneous representation as $\lambda_p^i \mathbf{x}_p^i = \mathbf{P}^i \mathbf{X}_p$ where λ_p^i is a non-zero scale called *projective depth*. Projections of all points into all images can be gathered into one large matrix equation $\mathbf{M} = \mathbf{PX}$ where $\mathbf{M}$ is so called *rescaled measurement matrix* which contains images of all points rescaled by projective depths, $3m \times 4$ matrix $\mathbf{P}$ contains m camera matrices stacked on top of each other, and a $4 \times n$ matrix $\mathbf{X}$ contains n 3D points. $\mathbf{M}$ has one column per 3D point and three rows per camera. If some point is not visible in some camera, the corresponding entries in matrix $\mathbf{M}$ are unknown. Projective structure, $\mathbf{X}$, and motion, $\mathbf{P}$, can be found by factorizing this large matrix. We use a modification of method [3] which is able to deal with both occlusions (missing entries) and outliers. This is necessary since we want to be able to do full reconstructions of objects of any shape, and since there may always be outliers among the matched points. Note that the method does not put any restrictions on image order.

All inliners with respect to the epipolar geometry, i.e. the pair-wise matches obtained from the region-matching step, can be placed into the M-matrix. In the original method [3], the conflicting matches are simply ignored and outliers removed in a subsequent stage using trifocal tensors. However, this turned out not to work when there were many image pairs with no mutual overlap. There will always be some matches between these image pairs and (incorrect) epipolar geometries (EGs) will be estimated with usually just a few matches satisfying them. Still, the number of matches may be higher than in some other image pair with a correct EG but with only a few matches due to small image overlap. Therefore, discarding pairs with the number of matches falling below some threshold does not work. A simple greedy algorithm can overcome this difficulty: First, matches from the image pair with the most inliers with respect to the EG are loaded into the M-matrix. Next, matches from the pair with the second largest number of inliers are loaded etc. This guarantees that more reliable matches are used first. Each match is checked against the already loaded EGs and if it satisfies them, it is merged into the M-matrix. It turns out that many outliers and only a few inliers are discarded this way.

Camera Auto-calibration. The reconstruction obtained in the last step is projective. To upgrade the projective reconstruction to a metric one, the cameras are auto-calibrated using the image of the absolute dual quadric [4]. The constraints used for the calibration are square pixels and zero skew. When more information about the cameras is available, more constraints can be used. To improve the quality of the solution, the calibration is followed by bundle adjustment including a radial distortion model.

Radial Distortion Correction. Since real cameras deviate from the linear pinhole model, the images have to be corrected for radial distortion. This is done by unwarping the images using the radial distortion model estimated in the bundle adjustment step described above. We use the division model $r_p = \frac{r}{1+\lambda r^2}$, where r_p is the perfect (undistorted) radius (measured from the distortion center $\mathbf{x}_0$), r the distorted radius and λ the radial distortion parameter. See [5] for the properties of this model. The reasons for using this model are that it is simple, performs well and that its inverse has a simple closed form except for at $r_p = 0$.

Image Pair Rectification. Rectification is necessary for an efficient dense matching procedure. After radial distortion rectification described above, the image pairs that will be used for dense matching are rectified by applying homographies mapping the epipoles to infinity on the horizontal axis [4]. This approach does not work if the epipoles are inside the images or too close to the image borders. Therefore image pairs, for which this is true are excluded from dense matching in an automatic close-pair selection step just before the rectification. The area around the epipoles would not provide 3D reconstructions of good geometric accuracy anyway and linearly rectified image pairs are more suitable for sub-pixel disparity estimation because of the simplicity of the underlying model which makes the algorithm faster and numerically better posed.

Dense Matching. Dense matching is performed as a disparity search along epipolar lines using Confidently Stable Matching (CSM) [6]. This algorithm assumes that the ordering constraint holds. If it does not, the corresponding part of the scene is rejected in the respective image pair. The CSM was used because is has a very low mismatch rate [7] and is fast. The single important parameter to CSM, for which a default value cannot be used, is the disparity search range. We set this parameter to the range of the known disparities of the sparse MSER matches plus a fixed margin. A typical search range for the overview images in the St. Martin scene (the church) is ±100 pixels. The algorithm has two more parameters: α, used to reject insufficient signal-to-noise ratio image data and β, used for repetitive pattern rejection (see [6]). By construction of the CSM algorithm none of these is critical nor scene-dependent. These parameters are both set to default values.

The output from the matching algorithm is one disparity map per image pair admitted for dense matching (see Fig. 4). By least squares estimation using an affine distortion model the disparity maps are upgraded to sub-pixel resolution [8].

Fig. 4. The disparity map for the first two images in the second row in Fig. 2 and Point clouds for the St. Martin scene (a front and a top view). Only 2% of all points are shown

Point Cloud Reconstruction and Local Aggregation to Fish-Scales. From the disparity maps the corresponding 3D points are reconstructed. The union of the points from all disparity maps forms a dense point cloud (see Fig 4).

An efficient way of representing distributions of points is to use fish-scales [9]. Fish-scales are local covariance ellipsoids that are fit to the points. They can be visualized as small round discs (see the results in Figs. 5, 6 and 7). A collection of fish-scales approximate the spatial density function of the measurement in 3D space.

The most important parameter of the fish-scale fitting is the fish-scale size. A too small value results in a noisy and sparse model and a too large value does not model fine structures well. The appropriate fish-scale size is found by sampling the density of the point cloud in the neighborhood of a number of points and by grounding the fish-scale size on the median point density. Here we use one fish-scale size for modeling the overall structure of the object, and a smaller size for modeling the details reconstructed from the high-resolution images. One point cloud from the low and one point cloud from the high resolution images are reconstructed and fish-scales are fit to the two point clouds independently. The large fish-scales that are close to a small fish-scale are rejected, and the two results are then fused by their union.

Texturing. Texture can easily be mapped on the fish-scales. However, we first have to decide from which view to get the texture for a certain fish-scale. This is done by counting the number of points reconstructed from each image pair within a certain Mahalanobis distance from a fish-scale. The number of points per view (one view could be used in several image pairs) are counted and by taking the view with the highest number of points we get the image with the best resolution and best view of the fish-scale. The method for choosing the best view for texturing takes one parameter, the distance within which to count the points.

3   Experiments and Discussion

We have applied our 3D reconstruction method to three different scenes: the St. Martin rotunda at Vyšehrad in Prague, the sculpture "l'Ecoute" (Listening) by Henri de Miller in Paris, and the Valbonne church near Nice. Image sizes were about 1000×1400 pixels for the rotunda and the sculpture and 512×768 for the church. The used input images can be seen in Figures 2, 1, and 7.

The images of the St. Martin rotunda were specially acquired for the purpose of 3D reconstruction. Overall images capturing the whole or major parts of the building were taken in such a way that views from adjacent positions would have a reasonable overlap. From each shooting position a narrow baseline image pair was taken. The baselines were about 1 to 1.5 meters. From some of the positions zoomed-in images of areas with fine geometric structures or poor texture were taken. For the experiment presented in this article only a subset of the images taken were used (see Fig. 2). The Head images are not optimal, they form a simple semicircular sequence (see Fig. 1). The Valbonne images form two semicircular trajectories (see Fig.7).

For all scenes, the whole procedure was performed fully automatically and with the same parameters. The only prior knowledge used was which focal lengths were the same: For the Head scene we knew that the focal length and principal point were approximately the same for all images. For the St. Martin scene the focal length and principal point were the same within the narrow baseline pairs. This information was used when auto-calibrating the cameras. However, if this knowledge was not used, very similar results were obtained. No knowledge of internal camera parameters was available for the Valbonne scene.

The region matching could be done for all image pairs and no pairs had to be manually forbidden. Some matches were always found, although some image pairs had no overlap. However, these matches were quite few and our method for finding the cameras was able to deal with them.

The narrow-baseline pair selection was quite simple: First, all pairs not suitable for rectification are forbidden. Next, the image pair with the highest number of inliers from the MSER matching is chosen. After that the second best pair is added and so on. When choosing a new pair, we require that at least one of the images in the pair has not been used before. This way every image is matched to the best image possible. Although non-optimal, the method gave the desired result for the St. Martin scene plus one extra pair (7–9) and also a good result (1–2, 2–3, 3–4, 5–6, 7–8, 9–10) when applied to the Head scene images.

The dense matching was carried out as described above. At least 30 points per volume of fixed size was required to make a fish-scale. Results for the three scenes are shown in Figs. 5, 6 and 7.

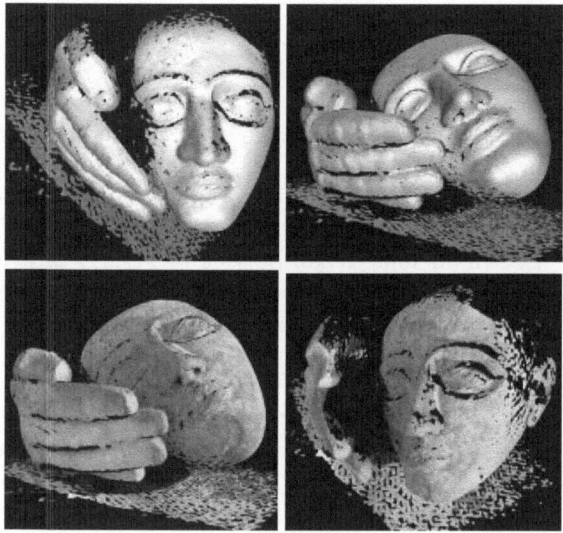

Fig. 5. The fish-scale model (textured and untextured) for the Head scene

Discussion. In all scenes, the models are smooth, and curvature is well captured. The reconstructions are highly accurate. For example, the pilaster on the apse of the rotunda is clearly visible in the reconstruction although it is only around 30 cm wide and less than 10 cm on the side and reconstructed from photographs taken approximately 20 meters away with a one meter baseline. Note also the thin structures like the ball at the cupola of the church. No jumps are visible on the boundaries between parts of the object reconstructed from different image pairs. The low-resolution fish-scales are aggregated from one single point cloud consisting of the points from all low resolution image pairs. Hence, any possible gaps along image borders would be smoothed-out. The blended-in details, on the other hand, are aggregated independently, from a different (denser)

Fig. 6. The fish-scale model for the St. Martin scene. Note the ball at the cupola and parts of the trees around the church. The first image in the second row is the top view. Texture is not radiometrically corrected to demonstrate which views contributed

point cloud, still no gaps are visible on borders between the two fish-scale sets, see the region around the door in Fig. 6.

The current version of fish-scale rendering has problems to capture thin and branching structures accurately, the rendering makes them more flat than in the actual model. This is visible on the ball at the cupola and on the branches of the surrounding trees.

In our approach we recover only spatial features that have *strong support in data*. Computational resources are not wasted to densely explain all—even weak-texture—data at once as in global optimization methods [10–12]. Fish-scales reconstructed from low-information patches would not contribute to the accuracy while increasing the demand for greater computational resources.

The fish-scales are considered as an intermediate 3D model. They capture local surface, including its orientation very efficiently. This makes them suitable for further camera parameter improvement, jointly with the 3D structure estimation. The result of such an iterative procedure could then be interpreted as a

Fig. 7. Input images and the fish-scale model for the Valbonne scene. Note that the walls of the tower are mutually perpendicular

triangulated surface as, for instance, in [13, 14, 9] or rendered directly as in [15]. Alternatively, the fish-scale model could be used as an initialization to a global optimization procedure that relies on high-accuracy camera calibration.

Our future work will include full-complexity fusion of partial reconstructions that requires selection of the best models on the partial model overlap. High-accuracy fish-scales should have precedence over low-accuracy fish-scales obtained from coarser-resolution images. Another part to be finished is bundle adjustment of the fish-scale model. In a final step we would like to densify or 'extend' the set of fish-scales by generating hypotheses around boundaries of the existing fish-scale sets and validate/refine them in the images, as in [16]. This possibility has been studied in [17]. We would also like to improve the close pair selection algorithm. In a large dataset it takes a very long time to run the MSER matching between all pairs of views. We study the possibility to avoid running the matching between all pairs. For example an approach similar to the one presented in [18] could be used.

4   Summary and Conclusions

In this work we have shown that given a wide baseline stereo matching algorithm, an occlusion-robust algorithm for estimating a consistent system of cameras from pair-wise point correspondences, and a dense stereo-matching algorithm, it is possible to obtain automatic high-resolution metric 3D reconstructions of objects of complex shape from a set of photographs. The strong requirement on the object to be reconstructed is that it must have sufficient texture, since this is required for the dense matching algorithm and for the geometric accuracy of the result. The requirements on the images are that among the photographs there have to be some image pairs suitable for dense matching and some wide baseline photographs to support the numerical stability of the camera calibration. By using image pairs of different resolutions it is possible to reconstruct the overall shape of the object from images with one resolution and to use higher resolution photographs for important details or poorly textured areas.

For the success of the data processing pipeline we find the following critical:

1. To obtain valid camera reconstructions there must be enough discriminatory regions (like the MSER) for the initial matching. This success is scene-dependent. Detecting sigle-type discriminatory regions need not suffice if the scenes are unconstrained in appearance.
2. The camera and scene reconstruction module has to cope with severe occlusion and moderate fraction of outliers in data.
3. Dense matching must produce few mismatches.
4. Surface reconstruction has to cope with complex structures, holes and missing data, and with a small to moderate fraction of outliers.

All points except for the first are satisfied in the method described here.

For the accuracy of the result the following is not critical but important: (1) Subpixel disparity estimation. (2) Radial distortion modeling. (3) Accurate epipolar geometry estimate before dense matching.

In this work we were surprised by the following: (1) The naroweness of the baseline for dense matching is not detrimental to the accuracy of the final model. (2) Higher resolution model parts did not require any additional effort to be blended in seamlessly. (3) It was possible to reconstruct thin structures consistently and with good accuracy, even from many uncalibrated views and even at the extremities of the scene like the ball at the cupola of the church. (4) Disparity maps need not be of full density to recover the fish-scale model because holes in one image pair are usually covered from another pair. (5) The method does not break if the cameras after the camera reconstruction step are inaccurate, as long as the epipolar geometry is accurate enough for the dense matching to work.

Acknowledgement. The authors would like to thank Jana Kostková for her help in data acquisition and Martin Matoušek for his implementation of the rectification procedure. Tomáš Werner provided the routine for the bundle adjustment. The Valbonne images were provided by courtesy of Andrew Zisserman. This project is supported by a grant from the STINT Foundation in Sweden under Project Dur IG2003-2 062, by the Czech Academy of Sciences under Project T101210406 and by IST Project IST-2001-39184 - BeNoGo.

References

1. Matas, J., Chum, O., Urban, M., Pajdla, T.: Robust wide baseline stereo from maximally stable extremal regions. In: BMVC. (2002) 384–393
2. Chum, O., Matas, J., Kittler, J.: Locally optimized RANSAC. In: DAGM. (2003) 236–243
3. Martinec, D., Pajdla, T.: Consistent multi-view reconstruction from epipolar geometries with outliers. In: SCIA. (2003) 493–500
4. Hartley, R., Zisserman, A.: Multiple View Geometry in Computer Vision. Cambridge University Press (2000)
5. Fitzgibbon, A.: Simultaneous linear estimation of multiple view geometry and lens distortion. In: Proc. CVPR. Volume 1. (2001) 125–132
6. Šára, R.: Finding the largest unambiguous component of stereo matching. In: ECCV. (2002) 900–914
7. Kostková, J., Čech, J., Šára, R.: Dense stereomatching algorithm performance for view prediction and structure reconstruction. In: SCIA. (2003) 101–107
8. Šára, R.: Accurate natural surface reconstruction from polynocular stereo. In: Proc NATO Adv Res Workshop Confluence of Computer Vision and Computer Graphics. Number 84 in NATO Science Series, Kluwer (2000) 69–86
9. Šára, R., Bajcsy, R.: Fish-scales: Representing fuzzy manifolds. In: ICCV. (1998) 811–817
10. Faugeras, O., Keriven, R.: Complete dense stereovision using level set method. In: ECCV. (1998) 379–393
11. Kutulakos, K.N., Seitz, S.M.: A theory of shape by shape carving. IJCV **38** (2000) 199–218
12. Kolmogorov, V., Zabih, R.: Multi-camera scene reconstruction via graph cuts. In: ECCV. (2002) 82–96
13. Han, S., Medioni, G.: Reconstructing free-form surfaces from sparse data. In: ICPR. (1996) 100–104
14. Amenta, N., Bern, M., Kamvysselis, M.: A new Voronoi-based surface reconstruction algorithm. In: SIGGRAPH. (1998) 415–421
15. Kalaiah, A., Varshney, A.: Modeling and rendering of points with local geometry. IEEE Trans on Visualization and Computer Graphics **9** (2003) 30–42
16. Ferrari, V., Tuytelaars, T., van Gool, L.: Wide-baseline multiple-view correspondences. In: CVPR. (2003) I: 718–725
17. Zýka, V., Šára, R.: Polynocular image set consistency for local model verification. In: OeAGM Workshop. (2000) 81–88
18. Schaffalitzky, F., Zisserman, A.: Multi-view matching for unordered image sets, or "How do I organize my holiday snaps?". In: ECCV. (2002) 414–431

Geometric Structure of Degeneracy for Multi-body Motion Segmentation

Yasuyuki Sugaya and Kenichi Kanatani

Depertment of Information Technology,
Okayama University, Okayama
700-8530 Japan
{kanatani, sugaya}@suri.it.okayama-u.ac.jp

Abstract. Many techniques have been proposed for segmenting feature point trajectories tracked through a video sequence into independent motions. It has been found, however, that methods that perform very well in simulations perform very poorly for real video sequences. This paper resolves this mystery by analyzing the geometric structure of the degeneracy of the motion model. This leads to a new segmentation algorithm: a multi-stage unsupervised learning scheme first using the degenerate motion model and then using the general 3-D motion model. We demonstrate by simulated and real video experiments that our method is superior to all existing methods in practical situations.

1   Introduction

Segmenting feature point trajectories tracked through a video sequence into independent motions is a first step of many video processing applications. Already, many techniques have been proposed for this task.

Costeira and Kanade [1] proposed a segmentation algorithm based on the shape interaction matrix. Gear [3] used the reduced row echelon form and graph matching. Ichimura [4] used the discrimination criterion of Otsu [13]. He also used the QR decomposition [5]. Inoue and Urahama [6] introduced fuzzy clustering. Kanatani [8, 9, 10] incorporated model selection by the geometric AIC [7] and robust estimation by LMedS [15]. Wu et al. [21] introduced orthogonal subspace decomposition.

According to our experiments, however, many methods that exhibit high accuracy in simulations perform very poorly for real video sequences. In this paper, we show that this inconsistency is caused by the *degeneracy* of the motion model on which the segmentation is based. The existence of such degeneracy was already pointed out by Costeira and Kanade [1]. Here, we report a new type of degeneracy, which we call *parallel 2-D plane degeneracy*, that, according to our experience, most frequently occurs in realistic scenes.

This discovery leads to a new segmentation algorithm, which we call *multi-stage unsupervised learning*: it operates first using our degeneracy model and then using the general motion model. We demonstrate that our method is superior to all existing methods in practical situations.

D. Comaniciu et al. (Eds.): SMVP 2004, LNCS 3247, pp. 13–25, 2004.

In Sec. 2, we describe the geometric constraints that underlie our method. In Sec. 3, we analyze the degeneracy of motion model. Sec. 4 describes our multi-stage learning scheme. In Sec. 5, we show synthetic and real video examples. Sec. 6 concludes this paper.

2   Geometric Constraints

Suppose we track N feature points over M frames. Let $(x_{\kappa\alpha}, y_{\kappa\alpha})$ be the coordinates of the αth point in the κth frame. Stacking all the coordinates vertically, we represent the entire trajectory by the following $2M$-D *trajectory vector*:

$$\boldsymbol{p}_\alpha = (x_{1\alpha} \ y_{1\alpha} \ x_{2\alpha} \ y_{2\alpha} \cdots x_{M\alpha} \ y_{M\alpha})^\top. \tag{1}$$

For convenience, we identify the frame number κ with "time" and refer to the κth frame as "time κ".

We regard the XYZ camera coordinate system as a reference, relative to which multiple objects are moving. Consider a 3-D coordinate system fixed to one moving object, and let $\boldsymbol{t}_\kappa$ and $\{\boldsymbol{i}_\kappa, \boldsymbol{j}_\kappa, \boldsymbol{k}_\kappa\}$ be, respectively, its origin and basis vectors at time κ. Let $(a_\alpha, b_\alpha, c_\alpha)$ be the coordinates of the αth point that belong to that object. Its position with respect to the reference frame at time κ is

$$\boldsymbol{r}_{\kappa\alpha} = \boldsymbol{t}_\kappa + a_\alpha \boldsymbol{i}_\kappa + b_\alpha \boldsymbol{j}_\kappa + c_\alpha \boldsymbol{k}_\kappa. \tag{2}$$

We assume an affine camera, which generalizes orthographic, weak perspective, and paraperspective projections [12, 14]: the 3-D point $\boldsymbol{r}_{\kappa\alpha}$ is projected onto the image position

$$\begin{pmatrix} x_{\kappa\alpha} \\ y_{\kappa\alpha} \end{pmatrix} = \boldsymbol{A}_\kappa \boldsymbol{r}_{\kappa\alpha} + \boldsymbol{b}_\kappa, \tag{3}$$

where $\boldsymbol{A}_\kappa$ and $\boldsymbol{b}_\kappa$ are, respectively, a 2×3 matrix and a 2-D vector determined by the position and orientation of the camera and its internal parameters at time κ. Substituting Eq. (2), we have

$$\begin{pmatrix} x_{\kappa\alpha} \\ y_{\kappa\alpha} \end{pmatrix} = \tilde{\boldsymbol{m}}_{0\kappa} + a_\alpha \tilde{\boldsymbol{m}}_{1\kappa} + b_\alpha \tilde{\boldsymbol{m}}_{2\kappa} + c_\alpha \tilde{\boldsymbol{m}}_{3\kappa}, \tag{4}$$

where $\tilde{\boldsymbol{m}}_{0\kappa}$, $\tilde{\boldsymbol{m}}_{1\kappa}$, $\tilde{\boldsymbol{m}}_{2\kappa}$, and $\tilde{\boldsymbol{m}}_{3\kappa}$ are 2-D vectors determined by the position and orientation of the camera and its internal parameters at time κ. From Eq. (4), the trajectory vector $\boldsymbol{p}_\alpha$ in Eq. (1) can be written in the form

$$\boldsymbol{p}_\alpha = \boldsymbol{m}_0 + a_\alpha \boldsymbol{m}_1 + b_\alpha \boldsymbol{m}_2 + c_\alpha \boldsymbol{m}_3, \tag{5}$$

where $\boldsymbol{m}_0$, $\boldsymbol{m}_1$, $\boldsymbol{m}_2$, and $\boldsymbol{m}_3$ are the $2M$-D vectors obtained by stacking $\tilde{\boldsymbol{m}}_{0\kappa}$, $\tilde{\boldsymbol{m}}_{1\kappa}$, $\tilde{\boldsymbol{m}}_{2\kappa}$, and $\tilde{\boldsymbol{m}}_{3\kappa}$ vertically over the M frames, respectively.

Eq. (5) implies that the trajectories of the feature points that belong to one object are constrained to be in the *4-D subspace* spanned by $\{\boldsymbol{m}_0, \boldsymbol{m}_1, \boldsymbol{m}_2, \boldsymbol{m}_3\}$ in $\mathcal{R}^{2M}$. It follows that multiple moving objects can be segmented into individual motions by separating the trajectories vectors $\{p_\alpha\}$ into distinct 4-D subspaces. This is the principle of the method of *subspace separation* [8, 9].

In addition, the coefficient of m_0 in Eq. (5) is identically 1 for all α. This means that the trajectories are in a *3-D affine space* within that 4-D subspace[1]. It follows that multiple moving objects can be segmented into individual motions by separating the trajectory vectors $\{p_\alpha\}$ into distinct 3-D affine spaces. This is the principle of the method of *affine space separation* [10].

Theoretically, the segmentation accuracy should be higher if we use stronger constraints. Indeed, it is reported that in simulation the affine space separation performs better than the subspace separation except in the case in which perspective effects are very strong and the noise is small [10]. For real video sequences, however, we have found that the affine space separation accuracy is often lower than that of the subspace separation [18, 19]. To resolve this inconsistency is the first goal of this paper.

3   Structure of Degeneracy

The motions we most frequently encounter are such that the objects and the background are translating and rotating 2-dimensionally in the image frame with varying sizes. For such a motion, we can choose the basis vector k_κ in Eq. (2) in the Z direction (the camera optical axis is identified with the Z-axis). Under the affine camera model, motions in the Z direction do not affect the projected image except for its size. Hence, the term $c_\alpha \tilde{m}_{3\kappa}$ in Eq. (4) vanishes; the scale changes of the projected image are absorbed by the scale changes of $\tilde{m}_{1\kappa}$ and $\tilde{m}_{2\kappa}$ over time κ.

It follows that the trajectory vector p_α in Eq. (5) belongs to the *2-D affine space* passing through m_0 and spanned by m_1 and m_2 [18, 19]. All existing segmentation methods based on the shape interaction matrix of Costeira and Kanade [1] assume that the trajectories of different motions belong to independent 3-D subspaces [8, 9]. Hence, degenerate motions cannot be correctly segmented.

If, in addition, the objects and the background do not rotate, we can fix the basis vectors i_κ and j_κ in Eq. (2) to be in the X and Y directions, respectively. Thus, the basis vectors i_κ and j_κ are common to all objects and the background, so the vectors m_1 and m_2 in Eq. (5) are also common. Hence, the 2-D affine spaces, or "planes", of all the motions are *parallel* (Fig. 1(a)).

Note that *parallel 2-D planes can be included in a 3-D affine space*. Since the affine space separation method attempts to segment the trajectories into different 3-D affine spaces, it does not work if the objects and the background undergo this type of degenerate motions. This explains why the accuracy of the affine space separation is not as high as expected for real video sequences.

---

[1] Customarily, m_0 is identified with the centroid of $\{p_\alpha\}$, and Eq. (5) is written as

$$\begin{pmatrix} p'_1 & \cdots & p'_N \end{pmatrix} = \begin{pmatrix} m_1 & m_2 & m_3 \end{pmatrix} \begin{pmatrix} a_1 & \cdots & a_N \\ b_1 & \cdots & b_N \\ c_1 & \cdots & c_N \end{pmatrix} \text{ or } W = MS, \text{ where } p'_\alpha = p_\alpha - m_0.$$

However, our formulation is more convenient for the subsequent analysis [12].

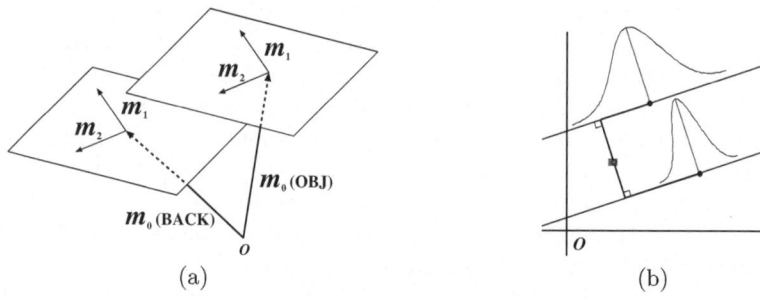

Fig. 1. (a) If the motions of the objects and the background are degenerate, their trajectory vectors belong to mutually parallel 2-D planes. (b) The data distributions inside the individual 2-D planes are modeled by Gaussian distributions

4   Degeneracy-Tuned Learning

We now define an unsupervised learning scheme [16] tuned to the parallel 2-D plane degeneracy. We assume that the noise in the coordinates of the feature points is an independent Gaussian random variable of mean 0 and a constant variance. We also model the data distributions inside the individual 2-D planes by Gaussian distributions (Fig.1(b)).

Let $n = 2M$. Suppose N n-dimensional trajectory vectors $\{p_\alpha\}$ are already classified into m classes by some means. Initially, we define the weight $W_\alpha^{(k)}$ of the vector p_α by

$$W_\alpha^{(k)} = \begin{cases} 1 & \text{if } p_\alpha \text{ belongs to class } k \\ 0 & \text{otherwise} \end{cases} \tag{6}$$

Then, we iterate the following procedures A, B, and C in turn until all the weights $\{W_\alpha^{(k)}\}$ converge[2].

A. Do the following computation for each class $k = 1, ..., m$.

1. Compute the fractional size $w^{(k)}$ and the centroid $p_C^{(k)}$ of the class k:

$$w^{(k)} = \frac{1}{N} \sum_{\alpha=1}^{N} W_\alpha^{(k)}, \qquad p_C^{(k)} = \frac{\sum_{\alpha=1}^{N} W_\alpha^{(k)} p_\alpha}{\sum_{\alpha=1}^{N} W_\alpha^{(k)}}. \tag{7}$$

2. Compute the $n \times n$ (second-order) moment matrix $M^{(k)}$:

$$M^{(k)} = \frac{\sum_{\alpha=1}^{N} W_\alpha^{(k)} (p_\alpha - p_C^{(k)})(p_\alpha - p_C^{(k)})^\top}{\sum_{\alpha=1}^{N} W_\alpha^{(k)}}. \tag{8}$$

---

[2] We stopped the iterations when the increments in $W_\alpha^{(k)}$ are all smaller than 10^{-10}.

B. Do the following computation.

1. Compute the *total* $n \times n$ moment matrix

$$\boldsymbol{M} = \sum_{k=1}^{m} w^{(k)} \boldsymbol{M}^{(k)}. \tag{9}$$

2. Let $\lambda_1 \geq \lambda_2$ be the largest two eigenvalues of the matrix $\boldsymbol{M}$, and $\boldsymbol{u}_1$ and $\boldsymbol{u}_2$ the corresponding unit eigenvectors.

3. Compute the *common* $n \times n$ projection matrices ($\boldsymbol{I}$ denotes the $n \times n$ unit matrix):

$$\boldsymbol{P} = \sum_{i=1}^{2} \boldsymbol{u}_i \boldsymbol{u}_i^{\top}, \qquad \boldsymbol{P}_{\perp} = \boldsymbol{I} - \boldsymbol{P}. \tag{10}$$

4. Estimate the noise variance in the direction orthogonal to *all* the affine spaces by

$$\hat{\sigma}^2 = \max[\frac{\mathrm{tr}[\boldsymbol{P}_{\perp} \boldsymbol{M} \boldsymbol{P}_{\perp}]}{n-2}, \sigma^2], \tag{11}$$

where $\mathrm{tr}[\,\cdot\,]$ denotes the trace and σ is an estimate of the tracking accuracy[3].

5. Compute the $n \times n$ covariance matrix of the class k by

$$\boldsymbol{V}^{(k)} = \boldsymbol{P} \boldsymbol{M}^{(k)} \boldsymbol{P} + \hat{\sigma}^2 \boldsymbol{P}_{\perp}. \tag{12}$$

C. Do the following computation for each trajectory vector $\boldsymbol{p}_{\alpha}$, $\alpha = 1, \, ..., \, N$.

1. Compute the conditional likelihood $P(\alpha|k)$, $k = 1, \, ..., \, m$, by

$$P(\alpha|k) = \frac{e^{-(\boldsymbol{p}_{\alpha}-\boldsymbol{p}_C^{(k)}, \boldsymbol{V}^{(k)-1}(\boldsymbol{p}_{\alpha}-\boldsymbol{p}_C^{(k)}))/2}}{\sqrt{\det \boldsymbol{V}^{(k)}}}. \tag{13}$$

2. Recompute the weights $\{W_{\alpha}^{(k)}\}$, $k = 1, \, ..., \, m$, by

$$W_{\alpha}^{(k)} = \frac{w^{(k)} P(\alpha|k)}{\sum_{l=1}^{m} w^{(l)} P(\alpha|l)}. \tag{14}$$

After the iterations of A, B, and C have converged, the αth trajectory is classified into the class k that maximizes $W_{\alpha}^{(k)}$, $k = 1, \, ..., \, N$.

In the above iterations, we fit 2-D planes of the same orientation to all classes by computing the common basis vectors $\boldsymbol{u}_1$ and $\boldsymbol{u}_2$ from all the data. We also estimate a common outside noise variance from all the data. Regarding the fraction $w^{(k)}$ (the first of Eqs. (7)) as the *a priori probability* of the class k, we compute the probability[4] $P(\alpha|k)$ of the trajectory vector $\boldsymbol{p}_{\alpha}$ conditioned to be in the class k (Eq. (13)). Then, we apply *Bayes' theorem* (Eq. (14)) to compute the *a posterior probability* $W_{\alpha}^{(k)}$, according which all the trajectories are reclassified.

---

[3] The value $\sigma = 0.5$ (pixels) is suggested in [17] as a reasonable estimate.
[4] Multipliers independent of α and k are omitted. They cancel out in Eq. (14).

Note that $W_\alpha^{(k)}$ is generally a fraction, so one trajectory belongs to multiple classes with fractional weights until the final classification is made.

This type of learning[5] is widely used for clustering, and the likelihood is known to increase monotonously by iterations [16]. It is also well known, however, that the iterations are very likely to be trapped at a local maximum. So, correct segmentation cannot be obtained by this type of iterations alone unless we start from a very good initial value.

5   Multi-stage Learning

If we *know* that degeneracy exists, we can apply the above procedure for improving the segmentation. However, we do not know if degeneracy exists. If the trajectories were segmented into individual classes, we might detect degeneracy by checking the dimensions of the individual classes, but we cannot do correct segmentation unless we know whether or not degeneracy exists.

We resolve this difficulty by the following multi-stage learning. First, we use the affine space separation assuming 2-D affine spaces, which effectively assumes planar motions with varying sizes. For this, we use the Kanatani's affine space separation [10], which combines the shape interaction matrix of Costeira and Kanade [1], model selection by the geometric AIC [7], and robust estimation by LMedS [15]. segmentation by using the parallel plane degeneracy model, as described in the preceding section.

The resulting solution should be very accurate if such a degeneracy really exists. However, rotations may exist to some extent. So, we relax the constraint and optimize the solution again by using the general 3-D motion model. This is motivated by the fact that if the motions are really degenerate, the solution optimized by the degenerate model is *not affected* by the subsequent optimization, because the degenerate constraints also satisfy the general constraints.

In sum, our scheme consists of the following three stages:

1. Initial segmentation by the affine space separation using 2-D affine spaces.
2. Unsupervised learning using the parallel 2-D plane degeneracy model.
3. Unsupervised learning using the general 3-D motion model.

The last stage is similar to the second except that 3-D affine spaces are separately fitted to individual classes. The outside noise variance is also estimated separately for each class. The procedure goes as follows.

Initializing the weight $W_\alpha^{(k)}$ by Eq. (6), we iterate the following procedures A and B in turn until all $\{W_\alpha^{(k)}\}$ converge[6]:

---

[5] This scheme is often referred to as the *EM algorithm* [2], because the mathematical structure is the same as estimating parameters from "incomplete data" by maximizing the logarithmic likelihood marginalized by the posterior of the missing data specified by Bayes' theorem.

[6] The convergence condition is the same as in Sec. 4: see footnote 2.

A. Do the following computation for each class $k = 1, ..., m$.

1. Compute the fraction $w^{(k)}$ and the centroid $p_C^{(k)}$ by Eqs. (7).
2. Compute the $n \times n$ moment matrix $M^{(k)}$ by Eq. (8).
3. Let $\lambda_1 \geq \lambda_2 \geq \lambda_3$ be the largest three eigenvalues of the matrix $M^{(k)}$, and $u_1^{(k)}$, $u_2^{(k)}$, and $u_3^{(k)}$ the corresponding unit eigenvectors.
4. Compute the $n \times n$ projection matrices

$$P^{(k)} = \sum_{i=1}^{3} u_i^{(k)} u_i^{(k)\top}, \qquad P_\perp^{(k)} = I - P^{(k)}. \qquad (15)$$

5. Estimate the noise variance in the direction orthogonal to the affine space of the class k by

$$\hat{\sigma}_k^2 = \max[\frac{\mathrm{tr}[P_\perp^{(k)} M^{(k)} P_\perp^{(k)}]}{n - 3}, \sigma^2]. \qquad (16)$$

6. Compute the $n \times n$ covariance matrix of the class k by

$$V^{(k)} = P^{(k)} M^{(k)} P^{(k)} + \hat{\sigma}_k^2 P_\perp^{(k)}. \qquad (17)$$

B. Do the following computation for each trajectory vector p_α, $\alpha = 1, ..., N$.

1. Compute the conditional likelihood $P(\alpha|k)$, $k = 1, ..., m$, by Eq. (13).
2. Recompute the weights $W_\alpha^{(k)}$, $k = 1, ..., m$, by Eq. (14).

After the iterations of A and B have converged, p_α is classified into the class k that maximizes $W_\alpha^{(k)}$, $k = 1, ..., m$.

6 Other Issues

We assume that the number m of motions is specified by the user. For example, if a single object is moving in a static background, both moving relative to the camera, we have $m = 2$. Many studies have been done for estimating the number of motions automatically [1, 3, 6], but none of them seems successful enough. This is because the number of motions is *not well-defined* [9]: one moving object can also be viewed as multiple objects moving similarly, and there is no rational way to unify similarly moving objects into one *from motion information alone*, except using heuristic thresholds or ad-hoc criteria. If model selection such as the geometric AIC [7] and the geometric MDL [11] is used[7], the resulting number of motions depends on criteria as reported in [9]. In order to determine the number m of motions, one needs high-level processing using color, shape, and other information.

The feature point trajectories tracked through video frames are not necessarily correct, so we need to remove outliers. If the trajectories were segmented

---

[7] The program is available at: http://www.suri.it.okayama-u.ac.jp/e-program.html

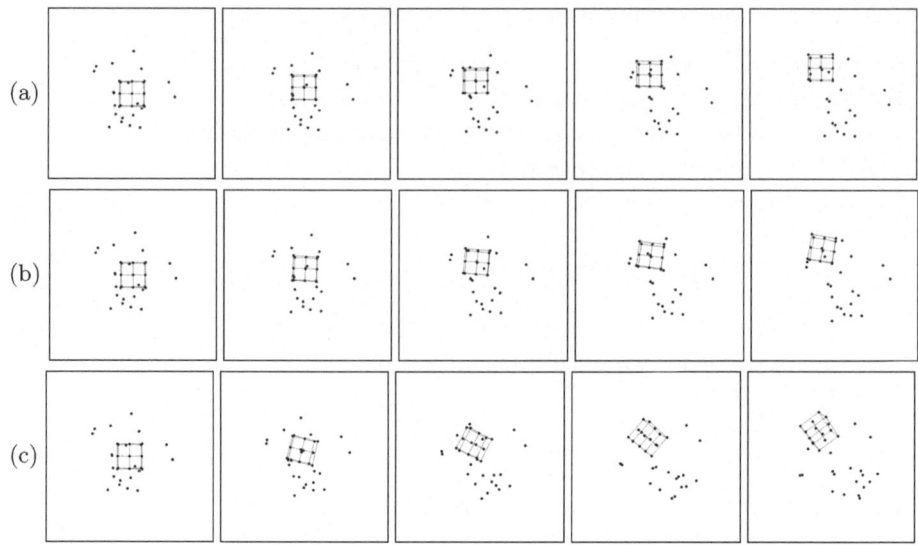

Fig. 2. Simulated image sequences of 14 object points and 20 background points: (a) almost degenerate motion; (b) nearly degenerate motion; (c) general 3-D motion

into individual classes, we could remove, for example, those that do not fit to the individual affine spaces. In the presence of outliers, however, we cannot do correct segmentation, and hence we do not know the affine spaces.

This difficulty can be resolved if we note that if the trajectory vectors $\{p_\alpha\}$ belong to m d-D subspaces, they should be constrained to be in a dm-D subspace and if they belong to m d-D affine spaces, they should be in a $((d+1)m-1)$-D affine space. So, we robustly fit a dm-D subspace or a $((d+1)m-1)$-D affine space to $\{p_\alpha\}$ by RANSAC and remove those that do not fit to it [17]. Thus, outliers can be removed *without knowing the segmentation results*. Theoretically, the resulting trajectories may not necessarily be all correct. However, we observed that all apparent outliers were removed by this method[8], although some inliers were also removed for safety [17].

7   Simulation Experiments

Fig. 2 shows three sequences of five synthetic images (supposedly of 512×512 pixels) of 14 object points and 20 background points; the object points are connected by line segments for the ease of visualization. To simulate real circumstances better, all the points are perspectively projected onto each frame with 30° angle of view, although the underlying theory is based on the affine camera model without perspective effects.

---

[8] The program is available at: http://www.suri.it.okayama-u.ac.jp/e-program.html

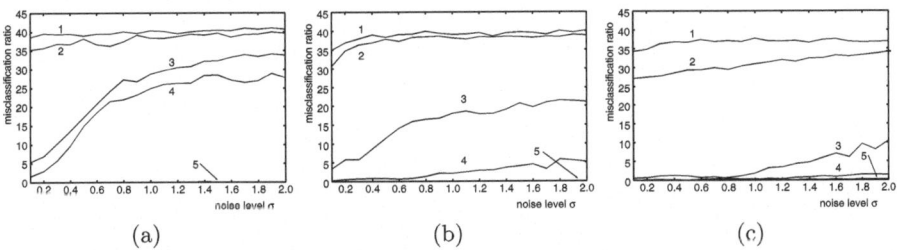

(a)   (b)   (c)

Fig. 3. Misclassification ratio for the sequences (a), (b), and (c) in Fig. 2: 1) Costeira-Kanade; 2) Ichimura; 3) optimized subspace separation; 4) optimized affine space separation; 5) multi-stage learning

In all the these sequences, the object moves toward the viewer in one direction ($10°$ from the image plane), while the background moves away from the viewer in another direction ($10°$ from the image plane). In (a), the object and the background are simply translating in different directions. In (b) and (c), they are additionally rotating by $2°$ per frame in opposite senses around different axes making $10°$ from the optical axis in (b) and $60°$ in (b). Thus, all the three motions are not strictly degenerate (with perspective effects), but the motion is almost degenerate in (a), nearly degenerate in (b), and a general 3-D motion in (c).

Adding independent Gaussian random noise of mean 0 and standard deviation σ to the coordinates of all the points, we segmented them into two groups. Fig. 3 plots the average misclassification ratio over 500 trials using different noise. We compared 1) the Costeira-Kanade method [1], 2) Ichimura's method [4], 3) the subspace separation [8, 9] followed by unsupervised learning (we call this *optimized subspace separation* for short), 4) the affine space separation [10] followed by unsupervised learning (*optimized affine space separation* for short), and 5) our multi-stage learning.

For the almost degenerate motion in Fig. 2(a), the optimized subspace and affine space separations do not work very well. Also, the latter is not superior to the former (Fig. 3(a)). Since our multi-stage learning is based on this type of degeneracy, it achieves 100% accuracy over all the noise range.

For the nearly degenerate motion in Fig. 2(b), the optimized subspace and affine space separations work fairly well (Fig. 3(b)). However, our method still attains almost 100% accuracy.

For the general 3-D motion in Fig. 2(c), the optimized subspace and affine space separations exhibit relatively high performance (Fig. 3(c)), but our method performs much better with nearly 100% accuracy again.

Although the same learning procedure is used in the end, the multi-stage learning performs better than the optimal affine space separation, because the former starts from a better initial value than the latter. For all the motions, the Costeira-Kanade method performs very poorly. The accuracy is not 100% even in the absence of noise ($\sigma = 0$) because of the perspective effects. Ichimura's method is not effective, either. It works to some extent for the general 3-D motion

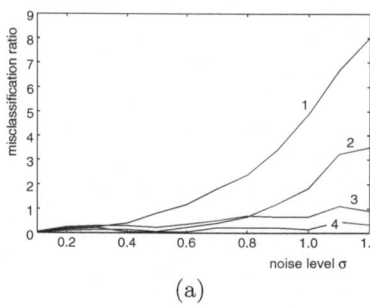
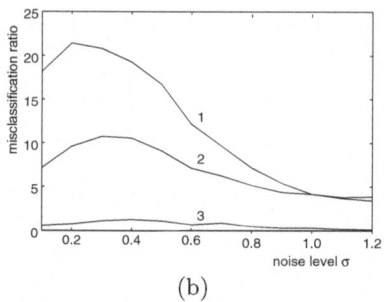

(a)   (b)

Fig. 4. Comparison of misclassification ratios: (a) Effects of unsupervised learning: 1) subspace separation; 2) optimized subspace separation; 3) affine space separation; 4) optimized affine space separation. (b) Stage-wise effect of multi-stage learning: 1) affine space separation using 2-D affine spaces; 2) unsupervised learning using the parallel 2-D plane degeneracy model; 3) unsupervised learning using the general 3-D motion model

in Fig. 2(c), but it does not compare with the optimized subspace or affine space separation, much less with our multi-stage optimization method.

Fig. 4(a) shows the effects of learning for Fig. 2(c). We can see that the learning works effectively. As compared with them, however, our multi-stage learning is far better. Fig. 4(b) shows the stage-wise effects of our multi-stage learning for Fig. 2(c). For this general 3-D motion, the learning using the parallel 2-D plane degeneracy model does not perform so very well indeed, but the subsequent learning based on the general 3-D motion model successfully restores the accuracy up to almost 100%. The interesting fact is that the accuracy increases as the noise increases. This reflects the characteristics of the initial affine space separation, whose accuracy deteriorates if perspective projection is affinely approximated for accurate data [10].

8   Real Video Experiments

Fig. 5 shows five decimated frames from three video sequences A, B, and C (320 × 240 pixels). For each sequence, we detected feature points in the initial frame and tracked them using the Kanade-Lucas-Tomasi algorithm [20]. The marks □ indicate their positions.

Table 1(a) lists the number of frames, the number of inlier trajectories, and the computation time for our multi-stage learning. The computation time is reduced by compressing the trajectory data into 8-D vectors [18]. We used Pentium 4 2.4GHz for the CPU with 1GB main memory and Linux for the OS. Table 1(b) lists the accuracies of different methods ("opt" stands for "optimized") measured by (the number of correctly classified points)/(the total number of points) in percentage. Except for the Costeira-Kanade and Ichimura methods, the percentage is averaged over 50 trials, since the subspace and affine space separations inter-

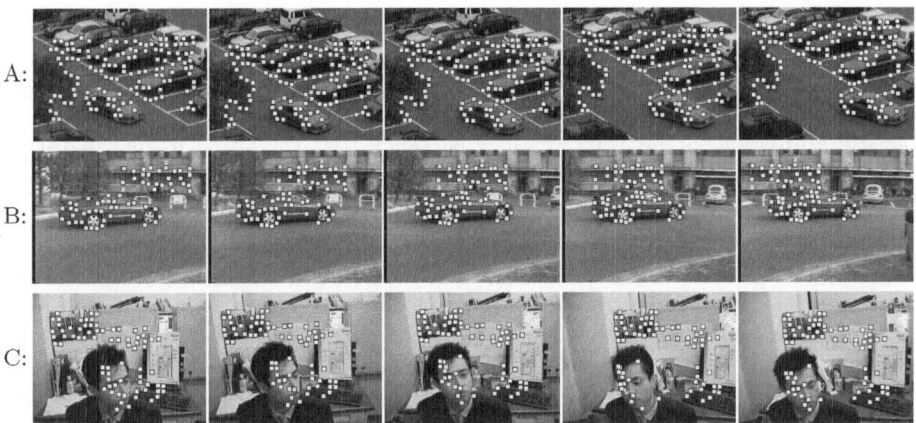

Fig. 5. Three video sequences and successfully tracked feature points

Table 1. (a) The computation time for the multi-stage learning of the sequences in Fig. 5. (b) Segmentation accuracy (%) for the sequences in Fig. 5

(a)

	A	B	C
# of frames	30	17	100
# of points	136	63	73
CPU time (sec)	2.50	0.51	1.49

(b)

	A	B	C
Costeira-Kanade	60.3	71.3	58.8
Ichimura	92.6	80.1	68.3
subspace separation	59.3	99.5	98.9
affine space separation	81.8	99.7	67.5
opt. subspace separation	99.0	99.6	99.6
opt. affine space separation	99.0	99.8	69.3
multi-stage learning	**100.0**	**100.0**	**100.0**

nally use random sampling for robust estimation and hence the result is slightly different for each trial.

As we can see, the Costeira-Kanade method fails to produce meaningful segmentation. Ichimura's method is effective for sequences A and B but not so effective for sequence C. For sequence A, the affine space separation is superior to the subspace separation. For sequence B, the two methods have almost the same performance. For sequence C, the subspace separation is superior to the affine space separation, suggesting that the motion in sequence C is nearly degenerate.

The effect of learning is larger for sequence A than for sequences B and C, for which the accuracy is already high before the learning. Thus, the effect of learning very much depends on the quality of the initial segmentation. For all the three sequences, our multi-stage learning achieves 100% accuracy.

9   Conclusions

In this paper, we analyzed the geometric structure of the degeneracy of the motion model that underlies the subspace and affine space separation methods [8–10] and resolved the apparent inconsistency that the affine space separation accuracy is often lower than that of the subspace separation for real video sequences. Our conclusion is that this is due to the occurrence of a special type of degeneracy, which we call *parallel 2-D plane degeneracy*.

Exploiting this finding, we proposed a multi-stage learning scheme first using the parallel 2-D plane degeneracy model and then using the general 3-D motion model. Doing simulations and real video experiments, we demonstrated that our method is superior to all existing methods in realistic circumstances.

The reason for this superiority is that our method is tuned to realistic circumstances, where the motions of objects and backgrounds are almost degenerate, whereas most existing methods implicitly assume that objects and backgrounds undergo general 3-D motions. As a result, they perform very poorly for simple motions such as in Fig. 5, while our method[9] has very high performance without compromising the accuracy for considerably non-degenerate motions.

References

1. J. P. Costeira and T. Kanade, A multibody factorization method for independently moving objects, *Int. J. Comput. Vision*, **29**-3 (1998-9), 159–179.
2. A. P. Dempster, N. M. Laird and D. B. Rubin, Maximum likelihood from incomplete data via the EM Algorithm, *J. Roy. Statist. Soc.*, B**39** (1977), 1–38.
3. C. W. Gear, Multibody grouping from motion images, *Int. J. Comput. Vision*, **29**-2 (1998-8/9), 133–150.
4. N. Ichimura, Motion segmentation based on factorization method and discriminant criterion, *Proc. 7th Int. Conf. Comput. Vision*, Vol. 1, Kerkyra, Greece, September 1999, pp. 600–605.
5. N. Ichimura, Motion segmentation using feature selection and subspace method based on shape space, *Proc. 15th Int. Conf. Pattern Recog.*, Vol. 3, Barcelona, Spain, September 2000, pp. 858–864.
6. K. Inoue and K. Urahama, Separation of multiple objects in motion images by clustering, *Proc. 8th Int. Conf. Comput. Vision*, Vol. 1, Vancouver, Canada, July 2001, pp. 219–224.
7. K. Kanatani, Geometric information criterion for model selection, *Int. J. Comput. Vision*, **26**-3 (1998-2/3), 171–189.
8. K. Kanatani, Motion segmentation by subspace separation and model selection, *Proc. 8th Int. Conf. Comput. Vision*, Vol. 2, Vancouver, Canada, July 2001, pp. 301–306.
9. K. Kanatani, Motion segmentation by subspace separation: Model selection and reliability evaluation, *Int. J. Image Graphics*, **2**-2 (2002-4), 179–197.
10. K. Kanatani, Evaluation and selection of models for motion segmentation, *Proc. 7th Euro. Conf. Comput. Vision*, Vol. 3, Copenhagen, Denmark, June 2002, pp. 335–349.

---

[9] The program is available at: http://www.suri.it.okayama-u.ac.jp/e-program.html

11. K. Kanatani, Uncertainty modeling and model selection for geometric inference, *IEEE Trans. Patt. Anal. Mach. Intell.*, **26**-10 (2004), to appear.

12. K. Kanatani and Y. Sugaya Factorization without factorization: Complete Recipe, *Memoirs of the Faculty of Engineering, Okayama University*, **38**-2 (2004), 61–72.

13. N. Otsu, A threshold selection method from gray-level histograms, *IEEE Trans. Sys. Man Cyber.*, **9**-1 (1979-1), 62–66.

14. C. J. Poelman and T. Kanade, A paraperspective factorization method for shape and motion recovery, *IEEE Trans. Pat. Anal. Mach. Intell.*, **19**-3 (1997-3), 206–218.

15. P. J. Rousseeuw and A. M. Leroy, *Robust Regression and Outlier Detection*, Wiley, New York, 1987.

16. M. I. Schlesinger and V. Hlaváč, *Ten Lectures on Statistical and Structural Pattern Recognition*, Kluwer, Dordrecht, The Netherlands, 2002.

17. Y. Sugaya and K. Kanatani, Outlier removal for motion tracking by subspace separation, *IEICE Trans. Inf. Syst.*, **E86-D**-6 (2003-6), 1095–1102.

18. Y. Sugaya and K. Kanatani, Automatic camera model selection for multibody motion segmentation, *Proc. Workshop on Science of Computer Vision*, Okayama, Japan, Sepember. 2002, pp. 31–39.

19. Y. Sugaya and K. Kanatani, Automatic camera model selection for multibody motion segmentation, *Proc. IAPR Workshop on Machine Vision Applications*, Nara, Japan, December 2002, pp. 412–415.

20. C. Tomasi and T. Kanade, *Detection and Tracking of Point Features*, CMU Tech. Rep. CMU-CS-91-132, Apr. 1991: http://vision.stanford.edu/~birch/klt/

21. Y. Wu, Z. Zhang, T. S. Huang and J. Y. Lin, Multibody grouping via orthogonal subspace decomposition, sequences under affine projection, *Proc. IEEE Conf. Computer Vision Pattern Recog.*, Vol. 2, Kauai, Hawaii, U.S.A., December 2001, pp. 695–701.

Virtual Visual Hulls: Example-Based 3D Shape Inference from Silhouettes

Kristen Grauman, Gregory Shakhnarovich, and Trevor Darrell

Computer Science and Artificial Intelligence Laboratory,
Massachusetts Institute of Technology,
Cambridge, MA 02139, USA
{kgrauman, gregory, trevor}@csail.mit.edu

Abstract. We present a method for estimating the 3D visual hull of an object from a known class given a single silhouette or sequence of silhouettes observed from an unknown viewpoint. A non-parametric density model of object shape is learned for the given object class by collecting multi-view silhouette examples from calibrated, though possibly varied, camera rigs. To infer a 3D shape from a single input silhouette, we search for 3D shapes which maximize the posterior given the observed contour. The input is matched to component single views of the multi-view training examples. A set of viewpoint-aligned virtual views are generated from the visual hulls corresponding to these examples. The most likely visual hull for the input is then found by interpolating between the contours of these aligned views. When the underlying shape is ambiguous given a single view silhouette, we produce multiple visual hull hypotheses; if a sequence of input images is available, a dynamic programming approach is applied to find the maximum likelihood path through the feasible hypotheses over time. We show results of our algorithm on real and synthetic images of people.

1   Introduction

Estimating the 3D shape of an object is an important vision problem, with numerous applications in areas such as virtual reality, image-based rendering, or view-invariant recognition. Visual hull methods, also called Shape-From-Silhouette (SFS), yield general and compact shape representations, approximating the 3D surface of an object by intersecting the viewing cones formed by the rays passing through the optical centers of a set of cameras and their corresponding image silhouettes. Typically a relatively small number of input views (4-8) is sufficient to produce a compelling 3D model that may be used to create virtual models of objects and people in the real world, or to render new images for view-dependent recognition algorithms.

In the absence of calibrated cameras, Structure-From-Motion (SFM) techniques may be used with a sequence of data to estimate both the observed object's shape as well as the motion of the camera observing it. Most such algorithms rely on establishing point or line correspondences between images and

D. Comaniciu et al. (Eds.): SMVP 2004, LNCS 3247, pp. 26–37, 2004.

frames, yet smooth surfaces without a prominent texture and wide-baseline cameras make correspondences difficult and unreliable to determine. Moreover, in the case of SFS, the occluding contours of the object are the only feature available to register the images. Current techniques for 3D reconstruction from silhouettes with an uncalibrated camera are constrained to the cases where the camera motion is of a known type, and the SFM methods cannot handle deformable, articulated objects.

In this paper we show that for shapes representing a particular object class, visual hulls (VHs) can be inferred from a single silhouette or sequence of silhouettes. Object class knowledge provides additional information about the object's structure and the covariate behavior of its multiple views. We develop a probabilistic method for estimating the VH of an object of a known class given only a single silhouette observed from an unknown viewpoint, with the object at an unknown orientation (and unknown articulated pose, in the case of non-rigid objects). We also develop a dynamic programming method for the case when sequential data is available, so that some ambiguities inherent in silhouettes may be eliminated by incorporating information revealed by how the object or camera moves.

We develop a non-parametric density model of the 3D shape of an object class based on many multi-view silhouette examples. The camera parameters corresponding to each multi-view training instance are known, but they are possibly different across instances. To infer a single novel silhouette's VH, we search for 3D shapes with maximal posterior probability given the observed contour. We use a nearest neighbor-based similarity search: examples which best match the contour in a single view are found in the database, and then the shape space around those examples is searched for the most likely underlying shape. Similarity between contours is measured with the Hausdorff distance. An efficient parallel implementation allows us to search 140,000 examples in a modest time.

To enable the search in a local neighborhood of examples, we introduce a new virtual view paradigm for interpolating between neighboring VH examples. Examples are re-rendered using a canonical set of virtual cameras; interpolation between 3D shapes is then a linear combination in this multi-view contour space. This technique allows combinations of VHs for which the source cameras vary in number and calibration parameters. The process is repeated to find multiple peaks in the posterior when the shape interpretation is ambiguous.

Our approach enables 3D surface approximation for a given object class with only a single silhouette view and requires no knowledge about either the object's orientation (or articulation), or the camera position. Our method can use sequential data to resolve ambiguities, or alternatively it can simply return a set of confidence-rated hypotheses (multiple peaks of the posterior) for a single frame. We base our non-parametric shape density model on the concise 3D descriptions that VHs provide: we can match the multi-view model in one viewpoint and then generate on demand the necessary virtual silhouette views from the training example's VH. Our method's ability to use multi-view examples from different camera rigs allows training data to be collected in a variety of real and synthetic environments.

2   Related Work

In this section we will review relevant related work on Shape-From-Silhouette algorithms and class-specific prior shape models.

Algorithms for computing the VH of an object have been developed based on the explicit geometric intersection of generalized cones [14]. Recent advances in VH construction techniques have included ways to reduce their computational complexity [16, 15], or to allow for weakly calibrated cameras [15]. A method combining SFS and stereo is given in [3] for the purpose of refining an object's VH by aligning its hulls from multiple frames over time. We rely on the efficient construction algorithm of [16] to calculate polygonal mesh VHs.

For the specific case of SFM using only the occluding contours, various methods have been devised which use knowledge of the uncalibrated camera's (or equivalently the rigid object's) motion to reconstruct surfaces. However, when applied to silhouette imagery current techniques are limited to certain types of camera motion (e.g., [4], [19], [21]).

When sequences of images are available, an alternative to geometric correspondence-based approaches like SFM is to utilize knowledge about the dynamics, or motion behavior, of the object moving in the video. For instance, knowledge about the types of motions the person is likely to perform may be exploited in order to infer the person's pose or shape. In the work of [2], a hidden Markov model is used to model the dynamics and 3D pose of a human figure in order to infer pose from a sequence of silhouettes by solving for the optimal path through the parametric model via entropy minimization. Our handling of sequential data uses dynamic programming to find the maximum likelihood path through a sequence of hypothesis virtual VHs. Our temporal integration step differs from that of [2] in that it processes different features (contours instead of central moments), and seeks to estimate a full 3D surface that fits the actual measurements of the input instead of rendering a cylinder-based skeleton from configural pose estimates.

A popular way to represent the variable shape of an object has been to employ a parametric distribution that captures the variation in the object shape. Such a model is often used for tracking, pose inference, or recognition. The use of linear manifolds estimated by PCA to represent an object class's shape, for instance, has been developed by several authors [13, 5, 1]. An implicitly 3D probabilistic shape model was introduced in [11], where a multi-view contour-based model using probabilistic PCA was given for the purpose of VH regularization. A method for estimating unknown 3D structure parameters with this model was given in [10]. However while [11, 10] require input views to be taken from cameras at the same relative angles as the training set, our method requires a single view with no calibration information at all.

Example-based and non-parametric density models of object shape have also been explored previously. In such models the object class is represented by a set of prototypical examples (or kernel functions centered on those examples), using either raw images or features extracted from them. For instance, the authors of [20] use 2D exemplars to track people and mouths in video sequences, a tem-

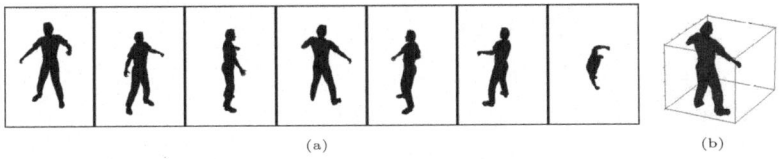

(a)   (b)

Fig. 1. (a) Examples in the model are composed of some number of silhouette views, plus their camera calibration parameters, which determine the VH (b)

plate hierarchy is developed for faster detection of pedestrian-shaped contours in [8], and in [17] a database of single view images with annotated body pose parameters is searched for a match to a test body shape based on its edges. In [18], a special mapping is learned from multi-view silhouettes to body pose. We employ a non-parametric density model constructed from many examples of multi-view silhouettes taken from different calibrated camera rigs.

We build a prior model of object shape using a non-parametric density model of multi-view contours. Given an observed single contour, we search for the shape most likely to have generated the observation. When the shape is ambiguous, we return multiple hypotheses corresponding to peaks in the posterior.

Our non-parametric density model for shape is defined by many multi-view silhouette instances of the object class, plus the associated camera calibration parameters (see Figure 1). Thus each example is equivalent to a VH. The number of cameras and their viewpoints may vary across examples, as long as for each example the camera parameters are recorded. The object's global orientation in each example is arbitrary. Generating a set of all possible examples to match discrete views is impractical, of course. A non-parametric density model gives us a principled way to interpolate between exemplars, which allows a relatively sparse set of examples to model a high dimensional space.

To measure the similarity between two contours A and B, represented by a set of uniformly sampled points, we use the Hausdorff distance:

$$||A - B||_H = \max(\max_{a \in A} D(a, B), \max_{b \in B} D(b, A)) , \qquad (1)$$

where $D(p, Q)$ is the shortest Euclidean distance from point p to any point in set Q. The Hausdorff distance has been proven to be an effective shape matching measure for such tasks as object recognition [12].

For the prior density over shapes we use a model of the form:

$$P(\mathbf{S}) = \frac{1}{Z} \sum_i^N K(\mathbf{S}, \mathbf{S}_i) , \qquad (2)$$

where Z is a normalizing constant, $\mathbf{S}$ is the 3D shape, and K is a *kernel* function defined in terms of the distance between shapes. We define this distance D_r in terms of the two shapes' rendered appearance over all viewpoints:

$$D_r(\mathbf{S}, \mathbf{S}_i) = \int ||\mathbf{s}^{\mathbf{P}} - \mathbf{s}_i{}^{\mathbf{P}}||_H d\mathbf{p} \,, \tag{3}$$

where $\mathbf{s}^{\mathbf{P}}$ is the rendering of shape $\mathbf{S}$ from a camera at pose $\mathbf{p}$. For VHs constructed from a finite set of views, we approximate D_r using a set of m fixed camera locations:

$$D_r(\mathbf{S}, \mathbf{S}_i) \approx \sum_{j=1}^{m} ||\mathbf{s}^{\mathbf{P}_j} - \mathbf{s}_i{}^{\mathbf{P}_j}||_H \,. \tag{4}$$

The training examples may be collected with real cameras, or generated synthetically (when the object class permits). The fact that the cameras in each example need not be the same is potentially a practical benefit when the model is built from real data, since this means that various camera rigs, at different locations, on different days, etc. may be employed to generate the examples.

2.1   3D Shape Inference with a Single Novel View

We will first explain the underlying method for estimating a single frame's VH. In fact, there could be several hypothesis estimates made at each frame in order to deal with ambiguous shapes; we discuss this process in Section 3.

To infer a VH from a single observed contour $\mathbf{C}$ in a Bayesian fashion, we wish to find $\mathbf{S}$ which maximizes

$$P(\mathbf{S}|\mathbf{C}) = \frac{P(\mathbf{C}|\mathbf{S})P(\mathbf{S})}{P(\mathbf{C})} \propto P(\mathbf{C}|\mathbf{S})P(\mathbf{S}) \,. \tag{5}$$

The observation likelihood is based on the similarity of a contour to any of the silhouette contours rendered from a given VH, which we approximate at a set of m discrete views at precomputed viewpoints:

$$P(\mathbf{C}|\mathbf{S}) = q \, \exp\left(-\min_{\mathbf{p}\in\Re^6} ||\mathbf{C} - \mathbf{s}^{\mathbf{P}}||_H^2/2\sigma^2\right) \approx \exp\left(-\min_{1<i<=m} ||\mathbf{C} - \mathbf{s}^{\mathbf{P}_i}||_H^2/2\sigma^2\right), \tag{6}$$

where q is a normalizing constant. Combining (2) and (6) we get the posterior that represents the most likely 3D shapes given an observed contour:

$$P(\mathbf{S}|\mathbf{C}) \propto \exp\left(-\min_{1<i<=m} ||\mathbf{C} - \mathbf{s}^{\mathbf{P}_i}||_H^2/2\sigma^2\right) \sum_{k=1}^{N} K\left(\sum_{j=1}^{m} ||\mathbf{s}^{\mathbf{P}_j} - \mathbf{s}_k^{\mathbf{P}_j}||_H\right). \tag{7}$$

In practice, one may use any kernel K which vanishes for sufficiently large values of its argument. We approximate the sum in the last term in (7) by the sum over the examples $\mathbf{S}_{(1)}, \ldots, \mathbf{S}_{(N)}$ for which $\left\|\mathbf{C} - \mathbf{S}_{(k)}^j\right\|_H \leq r$, for some view j. These r-*neighbors* of the input contour $\mathbf{C}$ can be found by means of thresholded similarity search in the database with respect to the Hausdorff distance[1].

---

[1] Similarity is measured on translation and scale invariant representations of the contours, which we obtain by subtracting the silhouette's 2D center of mass from each contour point's image coordinate and normalizing by its approximate size.

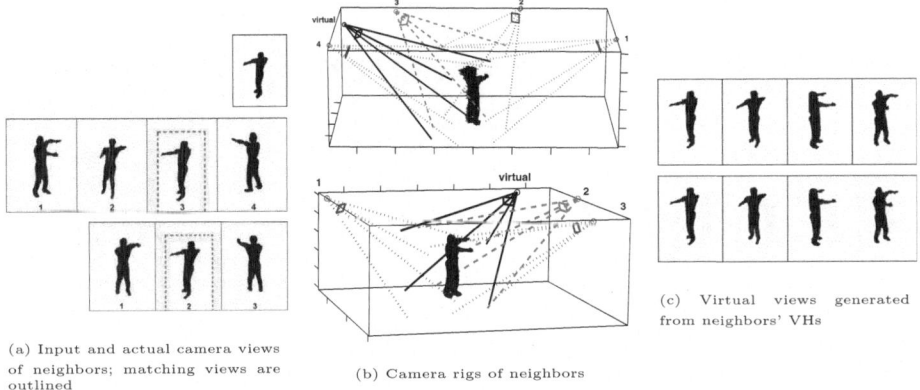

(a) Input and actual camera views of neighbors; matching views are outlined

(b) Camera rigs of neighbors

(c) Virtual views generated from neighbors' VHs

Fig. 2. An example of rendering canonical views from the r-neighbors of the input. Views that matched the input are marked with dotted boxes (a), and their corresponding cameras in the two examples' rigs are also shown with dotted lines (b). Viewpoint-aligned virtual views generated from the two neighbors' VHs are in (c)

A key problem is how to combine multiple 3D shape examples to form a single shape estimate; naive interpolation of unaligned triangle meshes will not yield meaningful shapes. Instead, we propose to interpolate between VHs through weighted combinations of the multi-view contours rendered from a set of canonical viewpoints, defined as follows. Each r-neighbor $\mathbf{S}$ of the contour $\mathbf{C}$ has a particular stored view $\mathbf{S}^j$ that best matches $\mathbf{C}$. We first refine an example's matching viewpoint by searching closely around its best stored matching viewpoint to find the view $\mathbf{S}^{j\prime}$ that even better matches the input silhouette in terms of contour distance. The first canonical viewpoint $\mathbf{p}_1$ for $\mathbf{S}$ corresponds to the viewpoint of the camera $j\prime$. The second canonical view $\mathbf{p}_2$ is obtained by a fixed transformation applied to $\mathbf{p}_1$ (say, rotation around the centroid of $\mathbf{S}$'s VH by the angle θ), etc. for v canonical viewpoints. This allows us to render, for each $\mathbf{S}$, v virtual silhouettes $\mathbf{r}^1, \ldots, \mathbf{r}^v$ so that every $\mathbf{r}^j$ corresponds to similar viewpoints (relative to the shape configuration) for all the neighbors of $\mathbf{C}$. Even though the real world position and orientation of the camera from each example which saw a view similar to the input may differ, each virtual view will be viewpoint-aligned across the neighbor examples from which they were generated.

To clarify with an example: suppose two multi-view neighbors $\mathbf{S}_{(1)}$ and $\mathbf{S}_{(2)}$ for a novel input contain four and three views with camera parameters $\{\mathbf{p}_1^1, \ldots, \mathbf{p}_4^1\}$ and $\{\mathbf{p}_1^2, \ldots, \mathbf{p}_3^2\}$, respectively (see Figure 2). Suppose the third view in the first example and the second view in the second example matched the input. Then the first virtual view for the first example is taken from the projection of its VH onto the image plane corresponding to the virtual camera found by rotating $\mathbf{p}_3^1$ by θ degrees about the VH's vertical axis. Similarly, the first virtual view for the second example is taken from a camera placed θ degrees from $\mathbf{p}_2^2$. Subsequent virtual views for each example are taken at equal intervals relative to

32   K. Grauman, G. Shakhnarovich, and T. Darrell

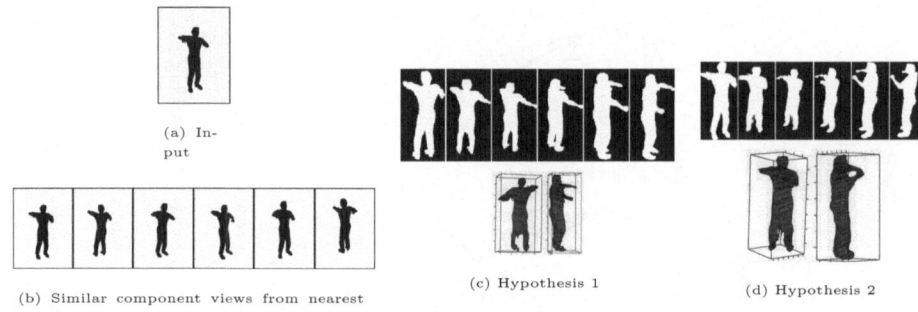

(a) In-
put

(b) Similar component views from nearest
neighbors

(c) Hypothesis 1

(d) Hypothesis 2

Fig. 3. An example where the input shape (a) has multiple interpretations. Its six nearest neighbors (b) contain views from examples originating from two general types of shapes: one front-facing body with the right elbow extended, the other back-facing with the left elbow extended. Aligned-viewpoint virtual views are projected from each of the six neighbors' VHs, and our algorithm finds two separate shape hypotheses (c,d) based on how these multi-view images cluster. Each hypothesis is shown from a frontal view and right side view, with the mean virtual views for each hypothesis shown above

these initial virtual cameras. After taking a weighted combination of the contours from these aligned-viewpoint virtual views, the output VH will be constructed using the camera parameters of the nearest neighbor's similar view and its virtual cameras; since it was the best match for the input in a single view, it is believed to contain a viewpoint relative to the object that is most like the true unknown input camera's, up to a scale factor.

After the set of canonical contours is produced as described above, they are normalized in location (translated) and length (resampled) in order to align the contour points in the same view across all the neighbors of $\mathbf{C}$ so that they may be combined, per view. We would like to obtain the resulting shape as a linear combination of the neighbors; recall, however, the previously stated objective of maximizing the *a posteriori* probability of the shape. Thus, we must find the vector of weights $\mathbf{w}$ such that

$$\mathbf{w}^* = \underset{\mathbf{w}}{\operatorname{argmax}} P\left(\sum_{k=1}^{N} w_k \mathbf{S}_{(k)} | \mathbf{C}\right) = \underset{\mathbf{w}}{\operatorname{argmax}} P\left(\mathbf{C} | \sum_{k=1}^{N} w_k \mathbf{S}_{(k)}\right) P\left(\sum_{k=1}^{N} w_k \mathbf{S}_{(k)}\right). \quad (8)$$

This can be done by means of gradient descent on the components of the vector $\mathbf{w}$. The shape hypothesis corresponding to the weights is then $\sum_k w_k^* \mathbf{S}_{(k)}$; note that this shape is, generally, no longer in the database, and may provide a better match for the input than any single training shape.

The 3D shape of a single view silhouette is inherently ambiguous: self-occlusions make it impossible to determine the full shape from a single frame, and the global orientation of the object may be uncertain if there is symmetry in the shape (e.g., a silhouette frontal view of a person standing with their legs side by side is similar to the view from behind). Thus we can expect the VHs corresponding to the "neighbor" single view silhouettes to manifest these different possible

3D interpretations. To combine widely varying contours from the neighbors' very different 3D shapes would produce a meaningless result (see Figure 3).

Instead, we maintain hypotheses corresponding to multiple peaks in the posterior at each time step. The nearest neighbors' aligned multi-view virtual silhouettes are clustered into enough groups such that the distance between two multi-view examples in one cluster is less than a threshold. Each cluster of examples will yield one peak of the posterior - one hypothesis VH. The single frame confidence in the VH hypothesis originating from a given cluster of neighbors is obtained by evaluating the posterior at the inferred VH for that cluster (see Eqn 7). Contours within the same cluster are topologically similar enough that they are expected to combine to form a valid shape. The hypotheses may be returned to a higher-level vision module, together with their confidences, or else they may be integrated using temporal data, as described below.

3   Integrating Single View Observations Over Time

When a sequence of observations is available, we apply a dynamic programming approach to find the maximum likelihood path through the feasible VHs at each time step, given our observations over time and the probabilities of transitioning from one shape to another. We construct a directed graph where each node is a time-indexed VH hypothesis (see Figure 4). Each directed path through the graph represents a legal sequence of states. Probabilities are assigned to each node and arc; node probabilities P_n^t are conditional likelihoods and arc probabilities P_a^t are transition probabilities:

$$P_n^t = P(\mathbf{C}^t | \mathbf{S}^t = \mathbf{S}_i)$$
$$P_a^t = P(\mathbf{S}^t = \mathbf{S}_j | \mathbf{S}^{t-1} = \mathbf{S}_i) \propto 1/D_r(\mathbf{S}_j, \mathbf{S}_i) \ . \tag{9}$$

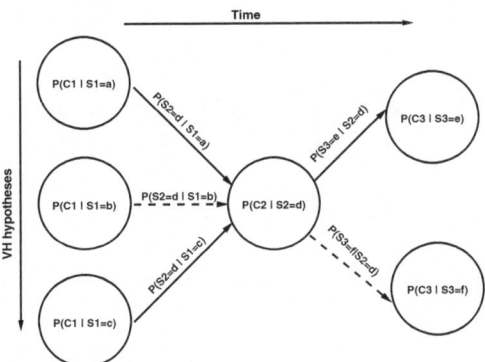

Fig. 4. Illustration of a directed graph corresponding to three consecutive frames. Nodes are VH hypotheses, edges are shape transition probabilities. The dotted line indicates the ML path found with dynamic programming

P_n^t is thus an estimate of the probability of observing contour $\mathbf{C}$ at time t given that the 3D shape of the object at time t is best approximated by the ith cluster hypothesis's VH, $\mathbf{S}_i$, and this is the evaluation of the likelihood (see Eqn 6). P_a^t is a measure of the similarity between two hypotheses at consecutive time steps, which we estimate in our experiments in terms of the sum of the Hausdorff distances between a set of canonical views rendered from the respective hypothesis VHs.

The maximum likelihood sequence of hypotheses in the graph is found using dynamic programming [7]. In this way an optimal path through the hypotheses is chosen, in that it constrains shapes to vary smoothly and favors hypotheses that were most likely given the observation. Note that this method for integrating temporal data remains general enough to handle different classes of objects, and requires no explicit model of dynamics for the object class. This process may be performed over windows of the data across time for longer sequences.

4 Experiments

We chose to build the model for human bodies from synthetic silhouettes using Poser, a computer graphics package [6]. The 3D person model is rendered from various viewpoints in randomized poses, and its silhouettes and camera parameters are recorded. An efficient parallel implementation allowed us to search 140,000 examples in modest time[2]. In our experiments we found the Hausdorff distance to be a robust measure of similarity between contours, since our input silhouettes were well-segmented, and we were able to reliably compute a scale and translation invariant representation of the contours.

We tested the inference method on both real and synthetic images. For the synthetic image tests, we generated a separate set of multi-view silhouettes using Poser, and then withheld each test example's VH information for ground truth comparisons. One view from each synthetic test example was input to our algorithm, and we measured the error E of the set of output VH hypotheses H for one test example as $E = \min_{h \in H} \left(\sum_{i=1}^{4} ||\hat{\mathbf{s}}^{\mathbf{P}_i} - \mathbf{s}^{\mathbf{P}_i}||_H \right)$, where $\hat{\mathbf{s}}^{\mathbf{P}_i}$ is the virtual

(a) Input view (left) and
two nearest neighbors

(b) Views from inferred VH

(c) Same views from actual VH

Fig. 5. Example of ground truth comparison for test on synthetic input with $E = 73$

[2] More recently we have designed a fast contour matching technique based on approximate nearest neighbors and the Earth Mover's Distance that allows us to use more elaborate shape descriptors and query a database of the same size in 1.5 seconds on a single processor [9].

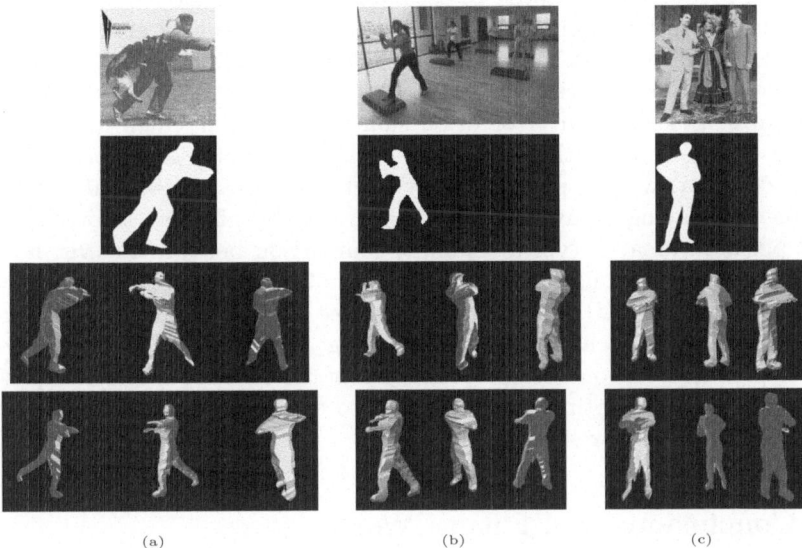

Fig. 6. Three example results on real single frame inputs. There are multiple hypothesis VHs due to the ambiguity in the single frame shapes (first two hypotheses are shown here for each example). Three viewpoints are rendered for each hypothesis

Fig. 7. Example result on real sequential data. Top row shows input sequence, middle row shows extracted silhouettes, and bottom row (output) shows VH hypotheses lying on the ML path, rendered here from a side view of the person in order to display the 3D quality of the estimate

view seen by a camera at pose $\mathbf{p}_i$ as inferred by our algorithm, and $\mathbf{s}^{\mathbf{p}_i}$ is the actual view seen by a camera at that pose for the withheld ground truth VH. For a synthetic set of 20 examples, the mean sum of Hausdorff distances over four views was 80. The Hausdorff distance errors are in the units of the image coordinates, thus a mean error of 80 summed over four views means that on average, per view, the farthest a contour point was from the ground truth contour (and vice versa) was 20 pixels. The synthetic test images are 240 x 320, with the contours having on average 800 points, and the silhouettes covering about 8,000 pixels in area. A typical example comparison between the virtual views generated by our estimated VH and the ground truth VH is shown in Figure 5.

We also inferred VHs for real images (see Figures 6 and 7). We consider these results to be preliminary but promising; the inferred VH hypotheses appear to provide a reasonable approximation of the 3D shape. The arm positions in Figure 7 are not optimal, and we plan to explore additional constraints on interpolation to improve the result.

5   Conclusions and Future Work

We developed a non-parametric prior model for a VH shape representation, and showed how 3D shape could be inferred from a single input silhouette or sequence of silhouettes with unknown camera viewpoints. Our prior model is learned by collecting multi-view silhouette examples from calibrated, though possibly varied, camera rigs. Visual hull inference consists of finding the shape hypotheses most likely to have generated the observed 2D contour. These peaks in the posterior are either returned directly, or, when a sequence of observations is available, integrated using a dynamic programming technique to find the most consistent trajectory of shapes over time.

Interpolation between neighboring examples allows our method to return shape estimates that are not literally in the set of examples used to define the prior model. We developed a new technique for 3D shape interpolation, using a set of viewpoint-aligned virtual views which are generated from the VHs corresponding to nearby examples. Interpolation between the contours of the aligned views produces a new set of silhouettes that are used to form the output VH approximating the 3D shape of the novel input.

We demonstrated our algorithm using a prior model trained with a large number of synthetic images of people. The accuracy of shape inference was evaluated quantitatively with held-out synthetic test images, and qualitatively with real images. We expect our method to be useful anytime a fast approximate 3D model must be acquired for a known object class, yet calibrated cameras or multiple cameras are not available. In the future we intend to explore ways to optimally combine the neighbors' contours, to deal with clutter in the shape matching stage, and to empirically study how the compactness of the database relates to the shape representational power of our method.

References

1. Baumberg, A., Hogg, D. An adaptive eigenshape model. BMVC, 1995.
2. Brand, M. Shadow puppetry. ICCV, 1999.
3. Cheung, G. K. M., Baker, S., Kanade, T. Shape-From-Silhouette of articulated objects and its use for human body kinematics estimation and motion capture. CVPR, 2003.
4. Cipolla, R., Blake, A. Surface shape from the deformation of apparent contours. IJCV 9(2) 1992.
5. Cootes, T., Taylor, C.A mixture model for representing shape variation.BMVC,1997.
6. Curious Labs, Egisys Co. Poser 5: The ultimate 3D character solution.
7. Forsyth, D., Ponce, J. Computer Vision: A Modern Approach. 2003. pp. 552–554.
8. Gavrila, D., Philomin, V. Real-time object detection for smart vehicles. ICCV 1999.
9. Grauman, K., Darrell, T. Fast contour matching using approximate earth mover's distance. CVPR, 2004.
10. Grauman, K., Shakhnarovich, G., Darrell, T. Inferring 3D structure with a statistical image-based shape model. ICCV, 2003.
11. Grauman, K., Shakhnarovich, G., Darrell, T. A Bayesian approach to image-based visual hull reconstruction. CVPR 2003.
12. Huttenlocher, D. Klanderman, G., and Rucklidge, W. Comparing images using the Hausdorff distance. PAMI, 1993.
13. Jones, M., Poggio, T. Multidimensional morphable models, ICCV, 1998.
14. Laurentini, A. The visual hull concept for silhouette-based image understanding. PAMI 16(2), 1994.
15. Lazebnik, S., Boyer, E., and Ponce, J. On computing exact visual hulls of solids bounded by smooth surfaces. CVPR, 2001.
16. Matusik, W., Buehler, C., McMillan, L. Polyhedral visual hulls for real-time rendering. EGWR, 2001.
17. Mori, G., Malik, J. Estimating human body configuration using shape context matching. ECCV, 2002.
18. Rosales, R., Sclaroff, S. Specialized mappings and the estimation of body pose from a single image. HUMO, 2000.
19. Szeliski, R., Weiss, R. Robust shape recovery from occluding contours using a linear smoother. IJCV 28(1), 1998.
20. Toyama, K., Blake, A. Probabilistic Exemplar-based tracking in a metric space. ICCV 2001.
21. Wong, K-Y. K. and Cipolla, R. Structure and motion from silhouettes. ICCV, 2001.

Unbiased Errors-In-Variables Estimation Using Generalized Eigensystem Analysis

Matthias Mühlich and Rudolf Mester

J.W. Goethe University, Frankfurt, Germany
[muehlich, mester]@iap.uni-frankfurt.de
http://www.uni-frankfurt.de/fb13/iap/cvg/

Abstract. Recent research provided several new and fast approaches for the class of parameter estimation problems that are common in computer vision. Incorporation of complex noise model (mostly in form of covariance matrices) into errors-in-variables or total least squares models led to a considerable improvement of existing algorithms.

However, most algorithms can only account for covariance of the same measurement – but many computer vision problems, e.g. gradient-based optical flow estimation, show correlations between different measurements.

In this paper, we will present a new method for improving TLS based estimation with suitably chosen weights and it will be shown how to compute them for general noise models. The new method is applicable to a wide class of problems which share the same mathematical core. For demonstration purposes, we have chosen ellipse fitting as a experimental example.

1   Introduction

1.1   The Errors-in-Variables (EIV) Model

Parameter estimation problems of the general form

$$\varphi(\boldsymbol{x}_{i0}, \boldsymbol{p}') = 0 \qquad \forall \quad i = 1, \ldots, m \tag{1}$$

are ubiquitous in computer vision. Here $\boldsymbol{p}' \in \mathbb{R}^n$ stands for the parameter vector that has to be estimated and $\boldsymbol{x}_{i0} \in \mathbb{R}^\ell$ denotes some true but unknown values (index i for different measurements), of which only some error-prone versions

$$\boldsymbol{x}_i = f(\boldsymbol{x}_{i0}, \boldsymbol{e}_i)$$

are available (for instance, $\boldsymbol{x}_i \in \mathbb{R}^4$ could be the stacked coordinates of corresponding points in a stereo image). Some (possibly non-linear) function f combines true values and errors; however, the assumption of additive noise $\boldsymbol{x}_i = \boldsymbol{x}_{i0} + \boldsymbol{e}_i$ is often reasonable. When $\boldsymbol{x}_{i0}$ is replaced by $\boldsymbol{x}_i$ in (1), we only achieve approximate equality: $\varphi(\boldsymbol{x}_{i0}, \boldsymbol{p}') \approx 0$.

Usually, we have an overdetermined system, i.e. (much) more measurements than unknown parameters, or, mathematically: $m > n$. The model defined by (1) is known as errors-in-variables (EIV) model.

D. Comaniciu et al. (Eds.): SMVP 2004, LNCS 3247, pp. 38–49, 2004.
© Springer-Verlag Berlin Heidelberg 2004

1.2 The Total Least Squares (TLS) Model

EIV estimation problems can be linearized to yield an equation

$$\boldsymbol{a'}_i^T \boldsymbol{p'} \approx b'_i \qquad (\boldsymbol{a'_i} \in \mathbb{R}^n) \tag{2}$$

for each measurement i. For many computer vision problems (most notably fundamental matrix estimation, homography estimation and camera calibration), this linearization means constructing bi-linear forms of $(\boldsymbol{x}^T, 1)^T$; in these cases, the common linearization scheme is known as direct linear transform (DLT) [1].

Stacking these row vectors on top of each other gives $\mathbf{A'}\boldsymbol{p'} \approx \boldsymbol{b'}$ with $\mathbf{A'} \in \mathbb{R}^{m \times n}$. For simplicity of notation, we add $\boldsymbol{b'}$ as an additional column to $\mathbf{A'}$: $\mathbf{A} = (\mathbf{A'}|\boldsymbol{b'})$. Analogously, we append -1 to $\boldsymbol{p'}$ to construct $\boldsymbol{p}$. We obtain the much more convenient homogeneous form:

$$\mathbf{A}\boldsymbol{p} \approx \boldsymbol{0} \, . \tag{3}$$

Estimation problems of this type are known as total least squares (TLS) problems[1].

1.3 Including Errors in the TLS Concept

A general error model errors is defined by $\mathbf{A} = f(\mathbf{A}_0, \mathbf{D})$ with a true data matrix $\mathbf{A}_0$ and an error matrix $\mathbf{D}$, both of which being unknown. The 'true' TLS equation $\mathbf{A}_0\boldsymbol{p} = \boldsymbol{0}$ only has a non-trivial solution if $\mathbf{A}_0$ is rank-deficient. Therefore, solving a TLS problem is equivalent to estimating a rank-deficient approximation $\hat{\mathbf{A}}$ for $\mathbf{A}$ such that

$$\|\mathbf{A} - \hat{\mathbf{A}}\| = \|\hat{\mathbf{D}}\| \quad \rightarrow \quad \min \tag{4}$$

under *some appropriate norm* (or *error metric*). Throughout this paper, quantities with a hat symbol on top denote estimated values.

The TLS solution is widely equated with the singular value of $\mathbf{A}$ corresponding to the smallest singular value. This narrows the view on the potential of TLS-based approaches considerably because nothing was said about some special error model yet and how to solve for it. With the only exception of the constraint between measurement and parameters being linearized (equation (1) vs equation (2)), this model is in no way less general than the EIV approach[2]. In this paper, the TLS notion will be restricted to the definitions given so far –

---

[1] In (1), we introduced a model with only one constraint per measurement. We set the multidimensional case aside to keep things simple here, but the extension to q constraints per measurement is straight forward: φ becomes vector-valued and the $\boldsymbol{a}_i$ and b_i in (2) have to be replaced by $q \times n$-matrices (resp. q-vectors) for each measurement. But the final result after stacking everything on top of each other is the same again: $\mathbf{A}\boldsymbol{p} \approx \boldsymbol{0}$.

[2] In most computer vision algorithms, even the usual assumption of additive noise in EIV problems carries over to the equally usual assumption of additive noise in TLS problems because elements of $\mathbf{A}$ are usually constructed as multi-linear forms of $\boldsymbol{x}_i$.

no assumption on certain error metrics is made with the term 'TLS' itself. It is important to stress that taking the right singular vector is just *one variant* of TLS-based methods; this method will be denoted plain TLS or PTLS from now on. Other TLS-based approaches can differ widely in the way they are solved; for instance, the constrained TLS (CTLS) [2] needs iterative optimization.

PTLS estimation is widely used because it provides a closed form solution that is very easy to compute – and under certain assumptions it is indeed a statistically optimal solution. If we use an additive error model $\mathbf{A} = \mathbf{A}_0 + \mathbf{D}$ and errors of all elements of $\mathbf{D}$ are zero-mean and independent and identically distributed (iid), then taking the right singular vector corresponding to the smallest singular value as TLS solution minimizes mean squared error of the estimate. If additionally errors are Gaussian, then PTLS is even a maximum likelihood estimator [3].

However, these assumption are often not very realistic; therefore, PTLS estimates can be very erroneous (e.g. highly biased in case of fundamental matrix estimation without prior data normalization). The reason is simple: PTLS implicitly takes the *Frobenius norm* as 'appropriate norm' in equation (4) – and of course, it is not always the 'appropriate norm' indeed.

Note that in general the iid noise assumption is even violated if the underlying measurements $\boldsymbol{x}_i$ contain iid noise because of the non-linearity of the constraints in (1). The linearization of this equation usually introduces data-dependent (heteroscedastic) error terms and the data normalization [1] that is common practice before applying PTLS effectively alleviates this effect.

1.4   TLS and Subspace Methods

The TLS estimation problem can easily be embedded in the more general class of *subspace estimation problems*. In these problems, we have to estimate a rank-deficient matrix from a noisy measured matrix of higher rank. A prominent computer vision problem that belongs to this group (and which is not a TLS problem) is the factorization method in multi-view structure from motion [4, 1].

Subspace problems can therefore be formulated as the problem of dividing a matrix $\mathbf{A} = \mathbf{A}_0 + \mathbf{D}$ into two parts $\mathbf{A} = \hat{\mathbf{A}} + \hat{\mathbf{D}}$ (*data subspace* and *error subspace*). The term 'subspace' refers to the row and column subspaces of the matrices $\hat{\mathbf{A}}$ and $\hat{\mathbf{D}}$.

The central point of this paper will be the presentation of the new scheme for subspace estimation problems (including TLS problems as a subset). This new approach called ETLS retains the property of providing an easy-to-compute closed-form solution. The rest of this paper is organized as follows: After a short review and comparison of different approaches to parameter estimation problems in computer vision in section 2, we will define a general error model for subspace problems in section 3. Then the equilibrated TLS (ETLS) approach will be introduced (section 4). We continue with some experimental results in section 5 before we conclude with the summary.

2   Literature Review

Different parameter estimation approaches differ in several aspects: computational complexity, robustness against outliers, ability to include uncertainty information on the measured quantities x_i, ability to account for correlations between different measurements x_i and x_j. Some methods provide confidence bounds or even full posterior probability distribution function for the estimates, some do not. Additionally, for iterative methods, the convergence properties can be meaningful.

Basically, the methods range from fast suboptimal approximations to exhaustive search on complex manifolds in high dimensional space and it is everything but easy to give a fair comparison. Nevertheless, we try to summarize the basic concepts used in computer vision problems, trying to focus on more general methods instead of highly specialized ones.

2.1   Complex Error Models and the TLS Model

The advantage of the TLS problem $a_i^T p \approx 0$ over the general EIV model $\varphi(x_i, p) \approx 0$ is that it is linear in the constraints. If we set possible problems with the linearization process aside [5], we can exploit the reduced mathematical complexity to allow a thorough statistical treatment of the errors, i.e. not only in terms of second order statistical moments, but compute likelihood functions $p(a_i|x)$ that only depend on the desired parameters.

In [6], Nestares et al. showed how to compute $p(a_i|x)$ if a probability density function for the noise $p(d_i)$ and a conditional prior on the nuisance parameters, i.e. $p(a_{i0}|x)$, are given. The likelihood function is defined by an integral over the nuisance parameters. For Gaussian noise and Gaussian priors with same covariance matrix (up to scale), this is possible analytically.

If noise and/or nuisance prior of measurement i are non-Gaussian (or if they are both Gaussian with different covariance matrices – the assumption of same covariances is highly restrictive!) one has to resort to numerical integration, but computation of the likelihood function still remains possible.

However, computation of the likelihood function for the whole matrix $\mathbf{A}$, i.e. $p(\mathbf{A}|x)$, is only tractable if the measurements are independent. Then $p(\mathbf{A}|x)$ can be written as a product of row-vector likelihoods $p(a_i|x)$. Under certain assumptions, it can be shown that the negative log-likelihood function is similar to the cost function defined for the HEIV model in [5].

2.2   Iterative Methods Based on Cost Functions

Many algorithms essentially consist in the optimization of a suitable cost function. The proper definition and minimization of such a function is can be a complicated task, especially if some information on the uncertainty of measurements x_i is to be incorporated. Most problems are only mathematically tractable if the uncertainty information is restricted to second order statistical moments, i.e. we assume that different covariance matrices for each measurement $\mathbf{C}_i = \mathsf{E}\left[\boldsymbol{x}_i \boldsymbol{x}_i^T\right]$ are given (under the assumption of zero-mean errors).

The (theoretically optimal) "gold standard" is the bundle-adjustment method derived in the photogrammetry community long ago. But the extreme computational complexity of this method is much more problematic in computer vision than in photogrammetry and this led to the development of several faster algorithms with comparable estimation quality.

An early problem of the subspace kind is ellipse fitting. For this problem, one famous early iterative algorithm was developed by Simpson. This algorithm, however, has the major drawback of producing biased estimates. Kanatani studied this problem and developed several renormalization schemes [7] which essentially consist in removing the estimated bias in each iteration step.

Chojnacki et al. developed an new and very straight forward iterative method based on a variational approach which is called fundamental numerical scheme (FNS) [8]. Recently, the FNS method was extended to incorporate some special kind of constraints on the estimated parameters (the constraint function $\phi(\boldsymbol{x}) = 0$ must be homogeneous of some degree κ; additionally, some conditions must apply for the measurement matrix) [9].

In [5], Matei and Meer provide an another iterative method for solving EIV problems when covariance matrices $\mathbf{C}_i$ are available for each measurement $\boldsymbol{x}_i$. This method is called heteroscedastic EIV or, in short, HEIV. This approach does not allow to handle ancillary constraints in the iterative estimation process, but the enforcement of these constraints by proper (iterative) projection on the manifolds defined by these constraints is studied.

Both HEIV and FNS do not allow the different measurements do be correlated. An important example for problems with strongly correlated measurements is gradient-based orientation estimation, e.g. the estimation of optical flow. For this problem, Ng et al gave a EIV-based solution in [10] which assumes simple iid errors in the images but handles the resulting (much more complicated!) noise model for the gradients correctly.

Summarizing this section, one can say that research during the last few years provided fast approximations to the bundle-adjustment algorithm which can handle heteroscedastic noise and show an estimation quality comparable to bundle-adjustment. General ancillary constraints, however, are usually hard to include in the algorithm itself, and (with certain exceptions for special forms of constraints) one has to resort to two step algorithms which consist of an unconstrained estimate followed by a subsequent enforcement of the constraints.

3   Generalized Error Model for Subspace Problems

In this section, we will define a generalized (additive) error model for the TLS problem $\mathbf{A} = \mathbf{A}_0 + \mathbf{D} \approx \mathbf{0}$ that abandons any reference to certain rows being constructed from certain independent measurements. Every element of $\mathbf{D}$ will be treated equally and arbitrary variances and covariances shall be allowed.

3.1   The Covariance Tensor of a Random Matrix

Let us assume that the error matrix $\mathbf{D}$ is a zero-mean random matrix and all covariances between elements ij and $k\ell$, i.e. $\mathsf{E}\left[(\mathbf{D})_{ij}(\mathbf{D})_{k\ell}\right]$, are known. We can now define a four-dimensional tensor $\mathcal{C} \in \mathbb{R}^{m \times n \times m \times n}$

$$(\mathcal{C})_{ijk\ell} = \mathsf{E}\left[(\mathbf{D})_{ij}(\mathbf{D})_{k\ell}\right]$$

which fully decribes the error structure up to second order statistics. This model is much more general than some other methods presented in the previous section where only different covariance matrices $\mathbf{C}_i$ for each row vector were allowed – in the framework presented here, this would mean $(\mathcal{C})_{ijk\ell} = (\mathbf{C}_i)_{j\ell}\,\delta_{ik}$. Sometimes this assumption is valid – but in many cases, e.g. for computation of optical flow from the structure tensor of the space-time volume, the row vectors are highly correlated.

3.2   Covariance Propagation for Tensors

If we start from the EIV approach (i.e. the measurement equation is not linearized yet), all we have is a $n \times \ell$ random matrix $\mathbf{E}$ containing errors in the ℓ-dimensional measurements. Analoguously to the linearized $\mathbf{D} \in \mathbb{R}^{n \times m}$, we can define a covariance tensor of $\mathbf{E}$ (denoted as $\mathcal{C}_E$).

Now the question arises of how to propagate the covariance information from $\mathcal{C}_E$ to $\mathcal{C}_D$, the covariance tensor of the TLS error matrix $\mathbf{D}$. But the answer is simple: We have to use the Jacobian matrix of the linearization functions, i.e. $\mathbf{J}^i = \frac{\partial a_i^T(\boldsymbol{x})}{\partial \boldsymbol{x}}$. This is best done row by row; the cross-covariance matrices of the rows i and j transform as:[3]

$$\mathsf{E}\left[d_i d_k^T\right] = \mathbf{J}^i \left(\mathsf{E}\left[e_i e_k^T\right]\right)(\mathbf{J}^k)^T \tag{5}$$

where d_i^T and e_i^T are the i-th row vectors of $\mathbf{D}$ and $\mathbf{E}$, respectively. This is a straight forward extention of covariance propagation for covariance matrices. Now that we know how to construct $\mathcal{C}_D$ if necessary, we will refer to this tensor as $\mathcal{C}$ (without subscript) for the rest of the paper.

3.3   Optimal Cost Function for Subspace Estimation Problems

The covariance tensor $\mathcal{C}$ can be rearranged to a symmetric and positive definite $(mn) \times (mn)$-matrix $\mathbf{C}$; this is exactly the covariance matrix of the 'vectorized' version $d \in \mathbb{R}^{mn}$ of the $m \times n$ error matrix $\mathbf{D}$ which can be constructed by stacking all column vectors on top of each other. Estimating the error matrix $\mathbf{D}$ is then equivalent to minimizing the Mahalanobis distance of the estimated $\hat{d}$:

$$J_{opt} = \hat{d}^T \mathbf{C}^{-1} \hat{d} \quad \rightarrow \quad \min \quad \text{subject to } \mathsf{rank}\{\mathbf{A} - \mathbf{D}\} = r$$

---

[3] If probability density functions of the errors e_i were available, it could be advisable to carry them through the full non-linearity instead and compute the second order statistical moments (i.e. covariances) afterwards. But in general, this leads to ugly integrals which have to be solved numerically.

In case of Gaussian errors in d (or in $\mathbf{D}$, respectively) it is a maximum likelihood estimate. We can also write down J_{opt} in matrix space ($\mathcal{C}^{-1}$ being the 'tensorized', i.e. simply rearranged[4], inverse covariance matrix $\mathbf{C}^{-1}$):

$$J_{opt} = \left\| \hat{\mathbf{D}} \right\|_{\mathcal{C}}^2 = \sum_{ijk\ell} (\mathbf{D})_{ij} (\mathcal{C}^{-1})_{ijk\ell} (\mathbf{D})_{k\ell} \qquad \text{subject to } \mathsf{rank}\{\mathbf{A} - \mathbf{D}\} = r \ . \quad (6)$$

This function J_{opt} defines the optimal matrix norm that should be used in the basic TLS minimization equation (4)[5]. This optimal cost function J_{opt} can be used both for iterative optimization and for evaluating the performance of different algorithms. In contrast to most (H)EIV- or FNS-based algorithms, minimizing J_{opt} gives the possibility to account for correlations between different measurements.

3.4   A New Decomposition of a Covariance Tensor

Let $\mathcal{C} \in \mathbb{R}^{m \times n \times m \times n}$ be the covariance tensor of the random matrix $\mathbf{D}$. By permuting second and third indices we can construct a $m \times m \times n \times n$ tensor that can be mapped on a matrix $\mathbf{M} \in \mathbb{R}^{m^2 \times n^2}$. This matrix $\mathbf{M}$ can be decomposed using the singular value decomposition, i.e.

$$\mathbf{M} = \sum_{p=1}^{n^2} \alpha_p \, \boldsymbol{x}_p \, \boldsymbol{y}_p^T$$

where $\boldsymbol{x}_p \in \mathbb{R}^{m^2}$ and $\boldsymbol{y}_p \in \mathbb{R}^{n^2}$ are the left resp. right singular vectors and σ_p are the singular values. Both $\boldsymbol{x}_p$ and $\boldsymbol{y}_p$ can be re-arranged to square matrices $\mathbf{X}_p \in \mathbb{R}^{m \times m}$ and $\mathbf{Y}_p \in \mathbb{R}^{n \times n}$. Doing this, the tensor $\mathcal{C}$ is decomposed into the following sum of basic tensors $\mathcal{T}_p$:

$$(\mathcal{C})_{ijk\ell} = \sum_{p=1}^{n^2} \underbrace{\alpha_p \, (\mathbf{X}_p)_{ik} \, (\mathbf{Y}_p)_{j\ell}}_{(\mathcal{T}_p)_{ijk\ell}} = \sum_{p=1}^{n^2} (\mathcal{T}_p)_{ijk\ell} \ . \quad (7)$$

An iid random matrix is defined by $(\mathcal{C})_{ijk\ell} = \delta_{ik}\delta_{j\ell}$. If this special covariance tensor is fed into the tensor decomposition algorithm described above, the result is

$$(\mathcal{C})_{ijk\ell} = (\mathbf{I}_m)_{ik} \, (\mathbf{I}_n)_{j\ell} \ , \quad (8)$$

i.e. the sum disappears and the only left and right 'singular matrices' are identity matrices. We can exploit the tensor decomposition to define a transformation

---

[4] In MATLAB, these rearrangement operations can easily be done by defining some tensor T and matrix M and setting T(:) = M(:); no special function is required.

[5] Note that the optimality of PTLS for iid noise can be seen easily from (6): For iid noise, $(\mathcal{C}^{-1})_{ijk\ell} = (\mathcal{C})_{ijk\ell} = \delta_{ik}\delta_{j\ell}$ and $J_{opt}^{\text{(iid-noise)}} = \sum_{ij}(\mathbf{D})_{ij}^2 = \|\mathbf{D}\|_F^2$ becomes identical to the Frobenius norm – which is exactly the norm under which rank approximations using SVD are optimal.

rule for covariance tensors. If a random matrix $\mathbf{D}$ is transformed according to $\hat{\mathbf{D}} = \mathbf{W}_L \mathbf{D} \mathbf{W}_R^T$, then its (decomposed) covariance tensor is transformed in the following way:

$$(\tilde{\mathcal{C}})_{ijk\ell} = \sum_p \alpha_p \, (\mathbf{W}_L \mathbf{X}_p \mathbf{W}_L^T)_{ik} \, (\mathbf{W}_R \mathbf{Y}_p \mathbf{W}_R^T)_{j\ell} \, . \tag{9}$$

In [11], Stewart introduced *cross-correlated matrices*. These are random matrices that can be constructed from an iid random matrix $\mathbf{D}$ by applying a transformation with arbitrary non-singular matrices $\mathbf{W}_L$ and $\mathbf{W}_R$. However, no method was provided to determine whether a given random matrix is cross-correlated or not. Using the calculus developed above, this is simple: Applying transformation rule (9) on (8) yields

$$(\tilde{\mathcal{C}})_{ijk\ell} = (\mathbf{W}_L \mathbf{W}_L^T)_{ik} \, (\mathbf{W}_R \mathbf{W}_R^T)_{j\ell} \, ,$$

i.e. cross-correlated matrices are exactly those matrices that are defined by one base tensor only. If the random matrix $\mathbf{D}$ given for a TLS problem is cross-correlated, then a whitening transformation can be computed; one only has to invert the Cholesky factors of $\mathbf{X}_1$ and $\mathbf{Y}_1$.

Some generalizations of basic subspace estimation algorithms for non-iid cases are in fact generalizations to cross-correlated noise. This applies to 2D homography estimation [12] and factorization with uncertainty information [4].

4   Equilibrated TLS

Under the error model $\mathbf{A} = \mathbf{A}_0 + \mathbf{D}$ with $\mathsf{E}\,[\mathbf{D}] = \mathbf{0}$, achieving unbiased subspace estimation boils down to two simple requirements that have to be fullfilled; this can be done with appropriate weighting transformations.

4.1   Unbiased Subspace Estimation Using Equilibration

In [13], TLS estimation is examined and it has been shown that unbiased estimates of right singular vectors of $\mathbf{A}$ (the smallest of which being the TLS solution vector) require that $\mathsf{E}\,[\mathbf{D}^T \mathbf{D}]$ is proportional to the identity matrix:

$$\mathsf{E}\,[\mathbf{D}^T \mathbf{D}] \propto \mathbf{I}_M \tag{10}$$

($\mathbf{I}_M$ denoting the $M \times M$ identity matrix). The reason for this is simple: right singular vectors are eigenvectors of $\mathbf{A}^T \mathbf{A} = \mathbf{A}_0^T \mathbf{A}_0 + \mathbf{A}_0^T \mathbf{D} + \mathbf{D}^T \mathbf{A}_0 + \mathbf{D}^T \mathbf{D}$. In expectation, the second and third term vanish and the eigenvectors of $\mathbf{A}^T \mathbf{A}$ are only identical to those of $\mathbf{A}_0^T \mathbf{A}_0$ (i.e. the true ones) if (10) holds; adding a multiple of the identity matrix only increases eigen*values*, but does not change eigen*vectors*.

If (10) does not hold, in general[6] the errors will introduce a bias in the right singular vectors. However, there is a very simple solution: transform the matrix to another (reweighted) space ($\tilde{\mathbf{A}} = \mathbf{W}_L \mathbf{A} \mathbf{W}_R$), do the rank reduction there, and transform back. This technique is called equilibration [14].

$$\mathbf{A} = \mathbf{W}_L^{-1} \underbrace{\mathbf{W}_L \mathbf{A} \mathbf{W}_R^T}_{\tilde{\mathbf{A}} = \mathbf{U} \mathbf{S} \mathbf{V}^T} \mathbf{W}_R^{-T} \tag{11}$$

Equilibration means that we approximate $\tilde{\mathbf{A}}$ with a rank-deficient matrix $\hat{\tilde{\mathbf{A}}}$. As long as both equilibration matrices are non-singular, they are guaranteed to preserve the rank; therefore transforming back gives an approximation $\hat{\mathbf{A}}$ for $\mathbf{A}$ with the desired lower rank.

It is obvious that a weighting transformation changes error metrics – but it was not clear *how* to choose weights in general, especially for the left equilibration. Fulfilling (10) in the transformed space is sufficient for unbiased TLS estimates. But in case of general subspace estimation it is not enough. SVD is defined by *both* the left and right singular vectors and not right singular vectors only. Even for TLS type estimates, a well chosen left equilibration is important because it reduces variance of the estimate.

An obvious extension of (10) is a second analogous requirement: $\mathsf{E}\left[\mathbf{D}\mathbf{D}^T\right] \propto \mathbf{I}$. In general (see footnote 6 again, now for both left and right singular vectors), fulfilling these two requirements *is the only way to get unbiased estimates for rank-deficient matrices* when using the SVD for matrix approximation.

Both requirements can be fulfilled in a transformed space, but the problem is the coupling we get for left and right equilibration matrices: the coupled equation system

$$\mathsf{E}\left[\tilde{\mathbf{D}}^T \tilde{\mathbf{D}}\right] = \mathbf{W}_R \mathsf{E}\left[\mathbf{D}^T (\mathbf{W}_L^T \mathbf{W}_L)\mathbf{D}\right] \mathbf{W}_R^T \propto \mathbf{I}_M \tag{12}$$

$$\mathsf{E}\left[\tilde{\mathbf{D}}\tilde{\mathbf{D}}^T\right] = \mathbf{W}_L \mathsf{E}\left[\mathbf{D}(\mathbf{W}_R^T \mathbf{W}_R)\mathbf{D}^T\right] \mathbf{W}_L^T \propto \mathbf{I}_N \tag{13}$$

cannot be solved for both $\mathbf{W}_L$ and $\mathbf{W}_R$ easily. But it becomes tractable with covariance tensor decomposition.

4.2   Computation of Equilibration Matrices

A reweighting of the measurement matrix according to (11), i.e. $\tilde{\mathbf{A}} = \mathbf{W}_L \mathbf{A} \mathbf{W}_R$, changes the covariance tensor $\mathcal{C}$ to

$$(\tilde{\mathcal{C}})_{ijk\ell} = \sum_p \alpha_p (\mathbf{W}_L \mathbf{X}_p \mathbf{W}_L^T)_{ik}(\mathbf{W}_R \mathbf{Y}_p \mathbf{W}_R^T)_{j\ell}$$

Here we see the advantage of the new tensor decomposition: The left (resp. right) hand equilibration matrix only affects the matrices $\mathbf{X}_p$ (resp. $\mathbf{Y}_p$).

---

[6] The degenerate case which preserves the correct eigenvectors is: $\mathbf{A}_0$ and $\mathbf{D}$ share the same right singular vectors and the adding of $\mathbf{D}$ does not change the order of singular values. In this very special case, (10) is not a necessary condition. Otherwise, it is.

The two requirements $\mathsf{E}\left[\tilde{\mathbf{D}}^T\tilde{\mathbf{D}}\right] \propto \mathbf{I}_N$ and $\mathsf{E}\left[\tilde{\mathbf{D}}\tilde{\mathbf{D}}^T\right] \propto \mathbf{I}_M$ now transform to

$$\sum_p \alpha_p \mathsf{Tr}\left\{\mathbf{W}_L\mathbf{X}_p\mathbf{W}_L^T\right\}(\mathbf{W}_R\mathbf{Y}_p\mathbf{W}_R^T) \propto \mathbf{I}_N \qquad (14)$$

$$\sum_p \alpha_p (\mathbf{W}_L\mathbf{X}_p\mathbf{W}_L^T)\mathsf{Tr}\left\{\mathbf{W}_R\mathbf{Y}_p\mathbf{W}_R^T\right\} \propto \mathbf{I}_M \ . \qquad (15)$$

This equation system (14) and (15) can be solved iteratively for $\mathbf{W}_L$ and $\mathbf{W}_R$, i.e. we

1. set $\mathbf{R}_{(0)} = 1/N\ \mathbf{I}_N$; $k = 1$
2. use (15) to compute $\mathbf{L}_{(k)}$ from $\mathbf{R}_{(k-1)}$ and scale it to Frobenius norm 1.
3. use (14) to compute $\mathbf{R}_{(k)}$ from $\mathbf{L}_{(k)}$ and scale it to Frobenius norm 1.
4. Terminate if $\mathbf{R}_{(k)}$ and $\mathbf{R}_{(k-1)}$ do not differ much; otherwise $k = k + 1$ and continue at step 2.

This procedure converges very fast and we never had any convergence problems. In case of cross-correlated matrices, both $\mathbf{W}_L$ and $\mathbf{W}_R$ are fully determined after the first iteration.

5   Experimental Results

The proposed method is applicable to a wide class of problems. We picked a very illustrative one for our experimental analysis: conic/ellipse fitting. This problem is important for many applications, known to be highly sensitive to errors and biased for PTLS estimation.

A conic in $\mathbb{R}^2$ is determined by a homogeneous 6-vector:

$$p_1 x^2 + p_2 y^2 + p_3\,xy + p_4\,x + p_5\,y + p_6 = 0$$

The measured data points are two-dimensional and the usual plain TLS approach suffers from the effect that a non-linear problem is embedded in a higher dimensional space which 'disturbs' the error metric. The result is well-known: if data points are only available from a certain arc of the ellipse, the fitted ellipses tend to be too small, i.e. the estimate is biased. We will show that our algorithm produces unbiased estimates.

We have defined a test ellipse (parameter vector $\boldsymbol{p} = (.1, .2, .5, 10, 10, 50)^T$) and taken 10 randomly chosen 2D points in the lower left arc of the ellipse. All points were disturbed by zero-mean additive Gaussian noise. The noise in the 10 measurements was *assumed to be correlated*, i.e. we used a random $10 \times 2 \times 10 \times 2$ covariance tensor (here: 190 degrees of freedom). Note that previously known ellipse fitting schemes like FNS [8] only allow individually different *but uncorrelated* covariance matrices for each measurement (here: $10 \cdot 3 = 30$ degrees of freedom).

From our perturbed data, we estimated the conic parameters (here: a unit vector in $\mathbb{R}^6$). One big advantage of ellipse fitting over other multidimensional parameter estimation problems is that the estimation quality can be visualized easily by plotting an ellipse with reconstructed parameters. Figure 1 shows 12 consecutive runs of our program. The solid light line is the true ellipse, the dashed line is the PTLS estimate and the dotted ellipse represents the ETLS estimate.

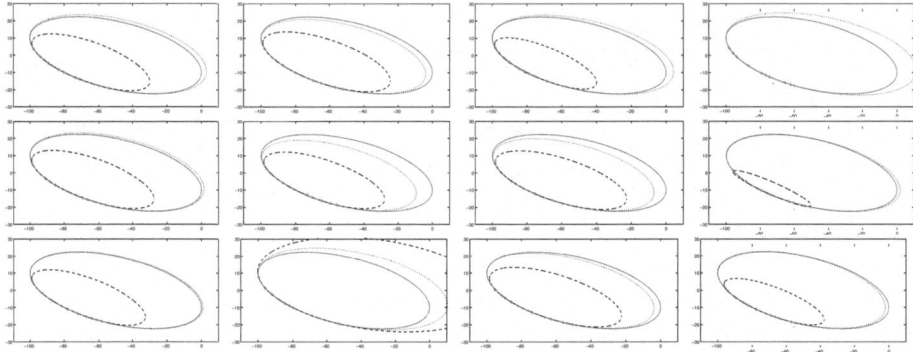

Fig. 1. 12 runs of our ellipse fitting algorithm. True ellipse (solid; identical in all images), PTLS estimate (dashed) and ETLS estimate (dotted). The bias of PTLS (ellipses tend to be too small) is clearly visible. Data points always chosen in lower left part of ellipse

The bias of the PTLS estimates is clearly visible in figure 1. In these 12 consecutive runs, the 'PTLS-ellipses' were too small 10 times, too large once, and in upper right image, the PTLS estimated conic was no ellipse at all.

We used a general conic estimation scheme here and no estimator which enforces an ellipse solution (e.g. [15]). Firstly, this leaves this example simple and easy to understand and secondly, it demonstrates that the necessity to prevent a wrong type of conic mainly arises by the usage of a wrong error metric. For the given example, our algorithm hardly ever estimates non-ellipses although it could in principle do so. Nevertheless, a combination of our approach and [15] is possible and should improve estimates in close situations.

6   Conclusion

Eigensystem based estimation schemes are ubiquitous in computer vision. Their estimation quality can often be improved by some previous data normalization, but still there is a desire to do further iterative optimization. The new method presented here provides a closed-form solution and might eliminate this need for many computer vision problems (or at least alleviate convergence for subsequent optimization).

Additionally, it is applicable in cases where correlation between different measurements has to be modelled; these problems cannot be treated properly with many other approaches which only account for covariances in the same measurement.

Ongoing research will focus on the development of a unified framework that combines the method presented here with approaches like HEIV and FNS; the mathematical core of most algorithms is very similar. The development of an optimal matrix norm for the TLS model is the first step in this direction.

References

1. Hartley, R.I., Zisserman, A.: Multiple View Geometry in Computer Vision. First edn. Cambridge University Press (2000)
2. Abatzoglou, T., Mendel, J., Harada, G.: The constrained total least squares technique and its applications to harmonic superresolution. IEEE Transactions on Signal Processing **3** (1991) 1070–1087
3. van Huffel, S., Vandewalle, J.: The Total Least Squares problem: Computational aspects and analysis. SIAM (Society for Industrial and Applied Mathematics), Philadelphia (1991)
4. Irani, M., Anandan, P.: Factorization with uncertainty. In: Proc. ECCV 2000. (2000) 539–553
5. Matei, B., Meer, P.: A general method for errors-in-variables problems in computer vision. In: IEEE Conference on Computer Vision and Pattern Recognition. (2000) 18–25
6. Nestares, O., Fleet, D.J., Heeger, D.J.: Likelihood functions and confidence bounds for Total Least Squares estimation. In: Proc. IEEE Conf. on Computer Vision and Pattern Recognition (CVPR'2000), Hilton Head (2000) 523–530
7. Kanatani, K.: Statistical Optimization for Geometric Computation: Theory and Practice. Elsevier (1996)
8. Chojnacki, W., Brooks, M.J., van den Hengel, A., Gawley, D.: On the fitting of surfaces to data with covariances. IEEE Transactions on Pattern Analysis and Machine Intelligence **22** (2000) 1294–1303
9. Chojnacki, W., Brooks, M.J., van den Hengel, A., Gawley, D.: A new approach to constrained parameter estimation applicable to some computer vision problems. In Suter, D., ed.: Proceedings of the Statistical Methods in Video Processing Workshop. (2002)
10. Ng, L., Solo, V.: Errors-in-variables modeling in optical flow estimation. IEEE Transactions on Image Processing **10** (2001) 1528–1540
11. Stewart, G.W.: Stochastic perturbation theory. SIAM Review **32** (1990) 576–610
12. Mühlich, M., Mester, R.: A considerable improvement in non-iterative homography estimation using TLS and equilibration. Pattern Recognition Letters **22** (2001)
13. Mühlich, M., Mester, R.: The role of total least squares in motion analysis. In: Proc. Europ. Conf. Comp. Vision. (1998)
14. Mühlich, M., Mester, R.: Subspace methods and equilibration in computer vision. Technical Report XP-TR-C-21, J.W.G.University Frankfurt (1999)
15. Halíř, R., Flusser, J.: Numerically stable direct least squares fitting of ellipses. In Skala, V., ed.: Proc. Int. Conf. in Central Europe on Computer Graphics, Visualization and Interactive Digital Media (WSCG98). (1998) 125–132

Probabilistic Tracking of the Soccer Ball

Kyuhyoung Choi and Yongdeuk Seo

Dept. of Media Technology, Sogang University,
Shinsu-dong 1, Mapo-gu, Seoul 121-742, Korea
{Kyu, Yndk}@sogang.ac.kr

Abstract. This paper proposes an algorithm for tracking the ball in a soccer video sequence. Two major issues in ball tracking are 1) the image portion of the ball in a frame is very small, having blurred white color, and 2) the interaction with players causes overlapping or occlusion and makes it almost impossible to detect the ball area in a frame or consecutive frames. The first is solved by accumulating the image measurements in time after removing the players' blobs. The resultant image plays a role of proposal density for generating random particles in particle filtering. The second problem makes the ball invisible for time periods. Our tracker then finds adjacent players, marks them as potential ball holders, and pursues them until a new accumulated measurement sufficient for the ball tracking comes out. The experiment shows a good performance on a pretty long soccer match sequence in spite of the ball being frequently occluded by players.

1   Introduction

Analysis of soccer video sequences has been an interesting application in computer vision and image analysis as can be seen by the abundance of recent papers let alone the fever of the soccer itself. Tracking players and ball must be a necessary step before an higher level analysis. There have been some researches on tracking players [1, 2, 3, 4, 5, 6, 7, 8, 9]. Among them, the papers such as [2, 3, 9] have dealt with the ball tracking problem as well. However ball tracking has not been thoroughly studied yet and that is the focus of this paper. Even though ball tracking belongs to single object tracking while player tracking falls within multi-object tracking, ball tracking is not easier than players tracking because of a few things. Usually ball blobs in images are very small, which makes it difficult to derive features from and to be characterized. Sudden changes in its motion is another factor to make it challenging. In addition, occlusion and overlapping with players causes a severe problem in tracking the ball continuously; The ball becomes invisible and appears at places where a continuous prediction could not reach. Yow *et. al.* [2] used intra-frame at regular intervals to detect the soccer ball without taking into account time continuity in the sequence. Seo *et. al.* [3] tracked the ball by template matching and Kalman filter, when there is no player close enough to the ball. When the ball is occluded and lost by nearby players, they searched for the ball in a bounding box around the players until a good

D. Comaniciu et al. (Eds.): SMVP 2004, LNCS 3247, pp. 50–60, 2004.

detection was obtained. In the work of Yamada *et. al.* [9], the nearest image blob in ground-free image to the predicted position by dynamics was considered as the ball among the candidates. The suggested ball tracking algorithm exploited player tracking results and a given background image to obtain images of the ball blob only and its trajectory.

The ball tracking as well as the players tracking in this paper is done by using particle filters, or equivalently, by SMC (Sequential Monte Carlo) methods [10, 11, 12, 13, 14]. In tracking multiple blobs of the players, we utilized the method proposed in [5] to address the problem of particle migration during occlusion between the same team players by probabilistic weighting of the likelihood of a particle according to the distance to its neighbors. This paper then concentrates on tracking the ball in a soccer video sequence. We utilize the result of players tracking in order to obtain measurement images that do not have players' blobs. Two major problems we consider in this paper are 1) the image portion of the ball in a frame is as small as 3×3 and the color is almost white but blurred due to its motion, and 2) the interaction with players causes overlapping or occlusion and makes it almost impossible to detect and predict the ball area in the sequence by a simple usage of a particle filter. To solve the first problem, we remove the image blobs of the players using the result of the players' tracking, segment out the ground field using a lower threshold, and finally accumulate the image blobs through the sequence. After an image filtering, this procedure results in a ball blobs connected continuously. Based on this accumulation image, particles are randomly generated only from those areas that have some blobs, which could be a noise blob, too, due to incomplete segmentation. Then, the particle filter evaluates each of the random particles to produce a tracking result. However, when occlusion or overlapping happens the accumulation does not provide meaningful ball blobs any more. In this case, our tracker changes the ball tracking mode to *invisible* from *visible*, finds and marks players near the location where the ball have disappeared, and chases the players instead of trying to estimate the ball location. This mode transition is done on the basis of the number of meaningful pixels in the accumulation image. For each player who is suspected (marked) to have the ball, searching for the ball is done in a pre-determined area with the player position as the center. When a player comes close enough to the marked, it also becomes enlisted. After a detection of the re-appearance of the ball by counting the meaningful pixels, the proposed algorithm resumes ball tracking.

Sequential Monte-Carlo method is explained in Section 2. Section 3 deals with pre-image processing and Section 3.1 covers player tracking. The method of ball tracking is discussed in 4. Section 5 provides experimental results and finally Section 6 concludes this paper.

2   Sequential Monte-Carlo Algorithm

Particle filtering or sequential Monte-Carlo (SMC) algorithm estimates the posterior distribution $p(x_t|z_t)$ sequentially, where x_t is the state and z_t is the measurement at time t, given a sequential dynamic equation with Gauss-Markov process. The posterior is represented by random particles or samples from the posterior distribution. When it is not possible to sample directly from the posterior distribution, a proposal distribution q of known random sampler can be adopted to compute the posterior, and in this case the posterior at time t is represented by the set of pairs of particle s and its weight w updated sequentially:

$$w_t = w_{t-1} \frac{p(z_t|x_t)p(x_t|x_{t-1})}{q(x_t|x_{0:t-1}, z_{1:t})} \tag{1}$$

After computation of w_t's for the particles generated from q and normalization $\sum_1^N w_t^i = 1$, where N is the number of particles, the set of particles comes to represent the posterior distribution. Particles have the same weight $1/N$ after re-sampling based on the weights or the posterior distribution.

Taking the proposal distribution as $q = p(x_t|z_{t-1})$ results in $w_t = w_{t-1} p(z_t|x_t)$, saying that the posterior can be estimated by evaluating the likelihoods at each time using the particles generated from the prediction process of system dynamics. Incorporated with resampling, the weight update equation can be further reduced to $w_t = p(x_t|z_t)$, where weight normalization is implied afterwards. This is the method of *condensation* algorithm [10, 11]. To solve the problem at hand by the condensation algorithm, one needs design appropriately the likelihood model $p(z|x)$ and state dynamic model $p(x|x_{t-1})$.

In this paper, the random proposal particles are not generated from $p(x|x_{t-1})$ in the ball tracking, but from a novel proposal distribution taking account of the accumulated measurements. Therefore, we use Equation 1 for updating the weights for the posterior density.

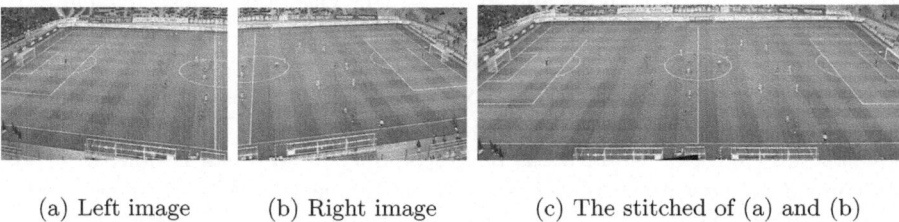

(a) Left image   (b) Right image   (c) The stitched of (a) and (b)

Fig. 1. Stitched image used as input to the tracking. The two video cameras are looking divergently, but placed so that their focal positions may be as close as possible

3   Pre-image Processing and Player Tracking

The source image to which we applied our tracking algorithm is a stitched version of two images. They are taken from the left and right camera mounted at the stadium respectively and formed into the one stitched image through the homography as shown in Figure 1. The two cameras are placed so that their focal positions may be as close as possible. The homography from one of the two to the stitched is computed by using points matches supplied manually.

(a) I^{bgd} (b) I_k^{ogn}

Fig. 2. Pre-processed images for players tracking. Left shows the constructed background image and right shows in input frame

The background image I^{bgd} is then obtained based on the methods [15, 16, 17]. The field part of original soccer image, I_k^{ogn} at frame k is subtracted to yield field-free image I_k^{sub} exploiting the background image I^{bgd}. Figure 2 shows the corresponding example images. In I_k^{sub}, the pixels of field parts are marked as black. Via CCL (connected component labeling) I_k^{ccl} is obtained. Size filtering deletes colored blobs that have either bigger or smaller enough size not to be considered as those of people.

3.1   Player Tracking

Player tracking is done in a similar way to [5] but using a different likelihood evaluation function. For the image I_k^{sub} of k th frame, state estimates of players $(\boldsymbol{p}_k)$ is done by the particle filter assigned respectively. The state vector of a player $\boldsymbol{p}$ is $(\boldsymbol{r}^T, w, h)^T$, where $\boldsymbol{r}$ is $(r_x, r_y)^T$ and represents the center position of a rectangle which a player is considered as. w and h mean the half width and height of the rectangle. Constant velocity is assumed for the dynamics of position and no velocity for the width and height.

$$\boldsymbol{r}_k = 2\boldsymbol{r}_{k-1} - \boldsymbol{r}_{k-2} \tag{2}$$

For particle filtering, each player has N samples or particles, the weight is determined by likelihood evaluation, that is, histogram comparison. A class is assigned to each player and has its model color histogram. When the color histogram of the region for the i-th ($i \in N$) sample s_A^i of player A is h_A^i and the model color histogram of the corresponding class is h_A, the likelihood L_i of s_A^i is expressed with total divergence D [18].

$$L_i = \frac{1}{\sqrt{2\pi}\sigma} \exp\left(\frac{-D(h_A, h_A^i)^2}{2\sigma^2}\right), \tag{3}$$

where

$$D(h_i, h_j) = \sum_{y \in B} \left\{ h_i(y) \log \frac{h_i(y)}{h_i(y) + h_j(y)} + h_j(y) \log \frac{h_j(y)}{h_i(y) + h_j(y)} \right\}. \tag{4}$$

Here y is an index of bin and B is the set of y given by:

$$B = \{y : h_i(y) > 0, h_j(y) > 0\} \tag{5}$$

The weights are obtained by normalizing the likelihoods L_i and the weighted sum of particles leads to $\hat{p}$, which is the estimate of p at this frame. The problem of particle migration during occlusion between the same team players is resolved by probabilistic weighting of the likelihood of particles according to the distance to its neighbors. See [5]. Some results of tracking multiple players are shown in 7 together with the results of ball tracking.

4   Ball Tracking

The basic idea in ball tracking is that the image consists of the players, ball and static background. So we may get I_t^{ball}, the image of ball only at the frame number t if we remove the portions of the background and players from the image. While player tracking is done at every single frame, ball tracking is batch processed at every m-th frame, where the interval of ball tracking is to produce a long enough accumulated area of the ball blobs. Examples of the accumulation are shown in Figure 3 and in our experiments the ball tracking interval m was 50 frames. If the blobs of players are deleted completely from the background-free image, we can get an accumulation image of I^{ball}s that is supposed to contain white pixels only from the ball area. However, notice that it contains noise pixels, too, due to incomplete background removal and players' blob detection. One could see that the ball has been in *visible* mode through the sequence since there are white accumulated areas (the linear structure in the accumulation image). The discontinuity means that the ball has been *invisible* during a period due to some reasons such as occlusion and overlapping.

The initial location of the ball is automatically detected by the proposed algorithm as following. If the interval m is 60, the accumulation image after the first interval is Figure 3(a) and it gives a filtered image of Figure 4(b) through CCL (See Figure 4(a)) and size filtering. If we have a filtered accumulation image the same as Figure 4(b) except that the non-black pixel value is the frame number instead of RGB color values, we can initialize the ball location, which is likely to be either end of the long blob in Figure 4(b), by finding pixels of minimum frame number. So during the first interval, accumulation images of ball blobs for both RGB color values and the frame number are made.

During the visible mode, we use a first order dynamic model for the ball motion perturbed by Gaussian random noise η:

$$x_t = 2x_{t-1} - x_{t-2} + \eta, \tag{6}$$

where $x = (x, y)$ is the location of the ball. The shape of the ball is modelled simply to be 3×3 rectangular. We measure the color values on the pixels in the 3×3 rectangle whose center is given by x - the state of the ball motion. Hence, our observation model for a ball particle is defined to be:

$$p(z_t|x_t) = \prod_i \prod_c \exp\left(-\frac{(c_i - \mu_c)^2}{\sigma_c^2}\right), \tag{7}$$

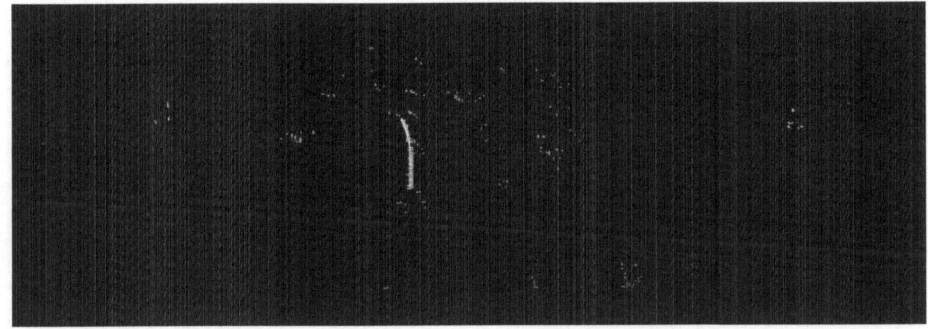

(a) After 60 frames

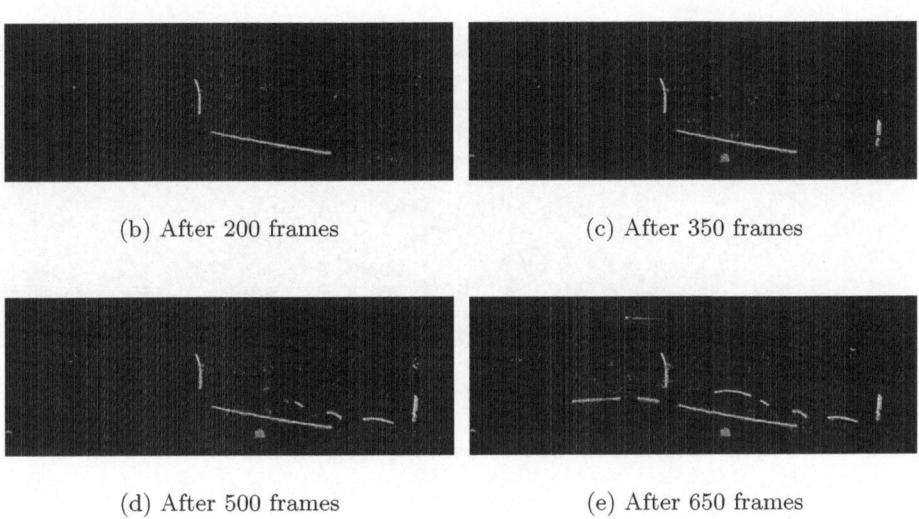

(b) After 200 frames   (c) After 350 frames

(d) After 500 frames   (e) After 650 frames

Fig. 3. Accumulation images for the ball blobs

where i denotes a pixel location i in the 3×3 rectangle, c_i the value in RGB color space at the pixel location, and μ_c and σ_c the mean and standard deviation calculated based on the pixel values around the ball area in a few video frames. Particles for the tracking is generated in the image region detected as the ball area after removing the players' blob and the background. Those pixels are designed to have equal probability and hence a uniform random sampler is utilized. The likelihood is evaluated using Equation 1, and the ball location is given by the weighted average of the particles.

When the ball is in the mode of *invisible*,we stop tracking the ball. In this case, the ball is assumed to be possessed by players near the place where the ball has become invisible. As shown in Figure 5, for each of the players who are suspected to have the ball, ball searching is done in the circled area with the player position as the center. Any player who comes close enough to the

(a) The connected component labelled image of Figure 3(a)

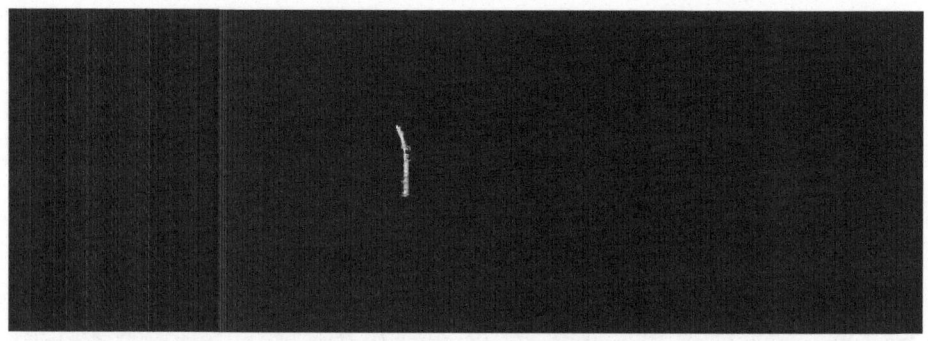

(b) The size-filtered image of Figure 3(a)

Fig. 4. Images in the steps of initializing the ball location

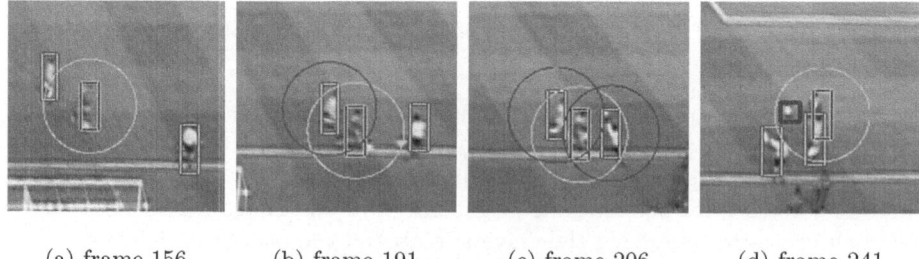

(a) frame 156   (b) frame 191   (c) frame 206   (d) frame 241

Fig. 5. Sub-images of some frames of interest

suspects also becomes enlisted. After the ball reappears and is detected through the accumulation, that is, one end of another ball blob trajectory (e.g. Figure 3) is found, the proposed algorithm resumes normal ball tracking as in the early part of this section.

In order to determine the ball tracking mode, we observe the number of pixels of the ball area in the accumulation image. At the frame number t ($t \neq 0$ and $(k-1)m \leq t < km$ for a natural number k), this value is given as the sum:

$$S_t = \sum_{j \in \{t-1, t, t+1\}} \sum_{l \in W_t} C_j(\boldsymbol{x}_l), \tag{8}$$

where $\boldsymbol{x}_l$ denotes an l-th pixel location in the search window W_t whose center is given by the estimated ball position at the frame number t, and C_j is an indication function:

$$C_j(\boldsymbol{x}) = \begin{cases} 0 & \text{if the color at } I_j^{ball}(\boldsymbol{x}) \text{ is black} \\ 1 & \text{otherwise} \end{cases} \tag{9}$$

Note that we incorporate the three consecutive image measurements in Equation 8 for a robust computation. Mode change is done simply by thresholding. When S_t is smaller than a threshold Th then the tracking mode changes to *invisible*, and as we explained before, the players are kept traced until our tracker finds the re-appearance of the ball pixels, that is, $S_t \geq Th$. Figure 6 shows a graph of $S_t - t$ for the input video of our experiments and S_t shows obvious changes at transitions between the two modes. At the most frames of *ivisible* mode S_t is zero and more than 50 for the *visible*. When $S_t < 15$ in the real

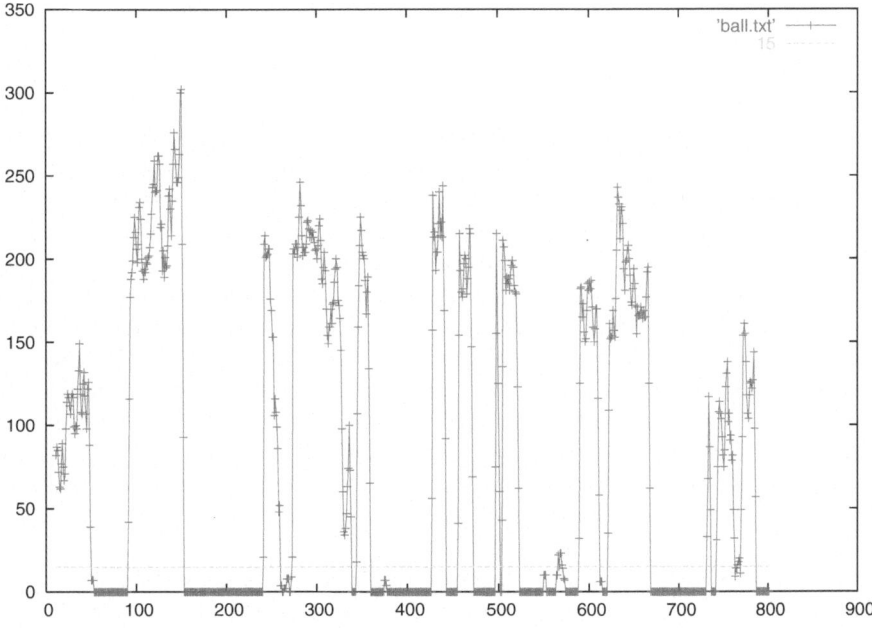

Fig. 6. Graph of $S_t - t$. Counting the number of meaningful pixels to determine the mode of the tracker

experiment, the mode changed to *invisible* and nearby players were traced to find the initiation of the ball blobs.

5   Experiments

Experiments were carried out on a video sequence of 800 images which are stitched as in Section 3. The image size was 1361×480 pixels. Figure 7 shows some frames of the results of which the detail is contained in accompanying video clip. The rectangle around each of the players is colored to show its identity. The inner color of its line stands for his class: ordinary players of each team, goal keeper of each team, and referee. A black circle around the ball means that the ball is not occupied by any player and thus the tracking mode is *visible*, and a colored circle shows the search area whose center is given by the location of the player, who is marked as a candidate having the ball. Notice that the color of the circle and the rectangle of the player are the same.

All the tracking process was done automatically without a manual intervention as well as the estimation of the initial ball location. The interval m was 60 and the threshold Th for the mode transition was set to $Th = 15$ which appears to be reasonable according to Figure 6. Note that even though players frequently repeat to keep, pass or kick the ball, the suggested algorithm showed a very robust performance and did not lose it, which is not guaranteed in other works such as [3, 9].

6   Conclusion

The algorithm presented in this paper have focused on an effective way of tracking the ball in a soccer match video sequence. The result of multiple player tracking was made use of in order to obtain a robust measurement for the ball tracking. By removing the blobs of players, we could obtain an accumulation image of the ball blobs. This accumulation image provided us not only a proposal density for the particle filtering but also a clue to deciding whether the ball was visible or invisible in the video frames. Basically, the ball tracking was done by particle filtering. However, the performance was highly improved by two ingredients: first, taking the accumulation image as the proposal density, and second, mode change by counting the meaningful ball pixels. When the ball was invisible, we pursued every nearby players until the ball pixel came out again. Since the ball pixels were accumulated in time, the tracking algorithm showed in the real experiment a very robust ball tracking results, that was not shown by other studies.

As a future work, we are going to develop a real time system for tracking through the whole game. Another interesting area is extending this work to the case of multiple camera sequences so that a full 3D reconstruction of the soccer game is possible.

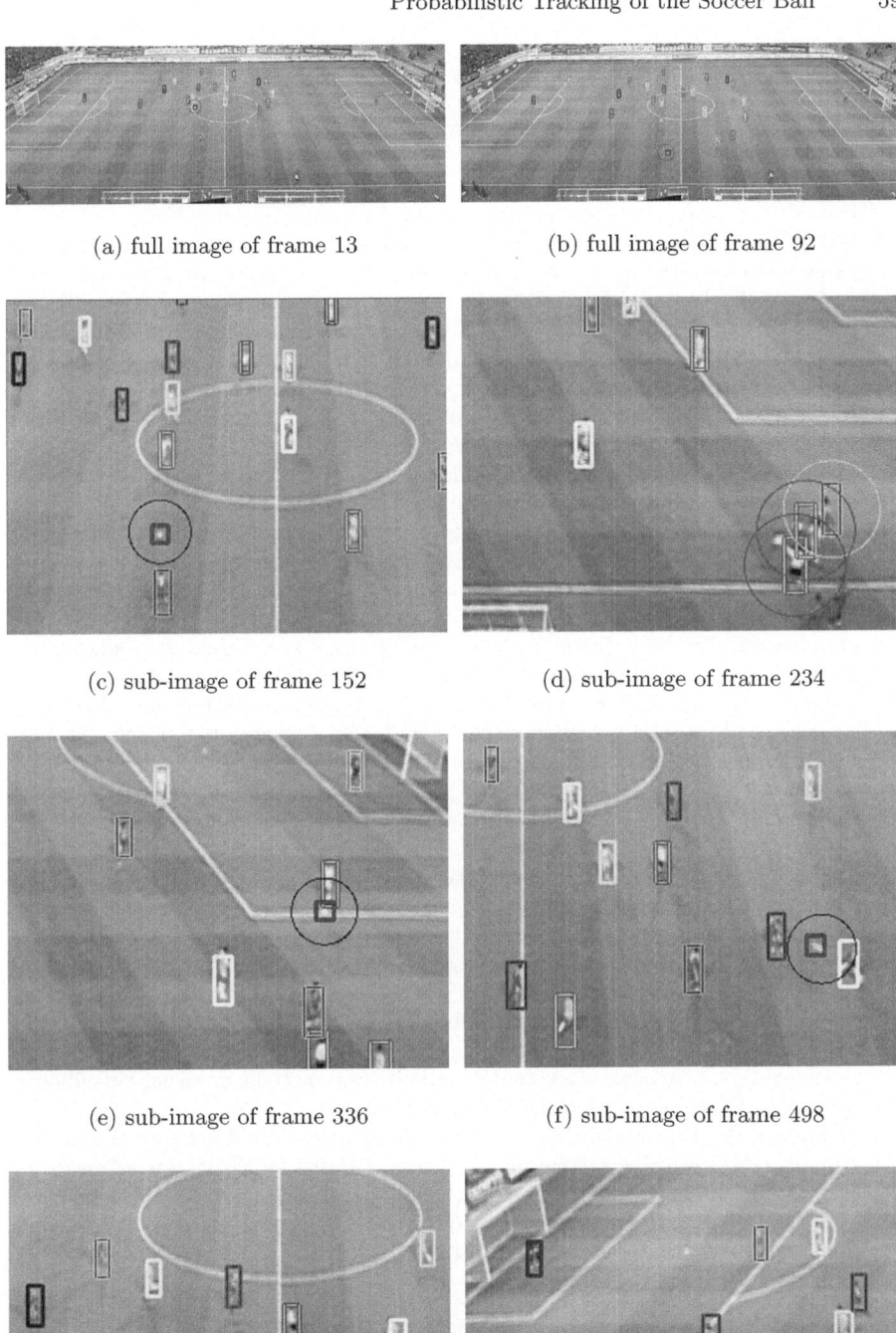

(a) full image of frame 13

(b) full image of frame 92

(c) sub-image of frame 152

(d) sub-image of frame 234

(e) sub-image of frame 336

(f) sub-image of frame 498

(g) sub-image of frame 588

(h) sub-image of frame 721

Fig. 7. Examples of result images

References

1. Intille, S., Bobick, A.: Closed-world tracking. In: Proc. Int. Conf. on Computer Vision. (1995)
2. Yow, D., Yeo, B., Yeung, M., Liu, B.: Analysis and presentation of soccer highlights from digital video. In: Proc. Asian Conf. on Computer Vision. (1995)
3. Seo, Y., Choi, S., Kim, H., Hong, K.: Where are the ball and players? soccer game analysis with color-based tracking and image mosaick. In: Proc. Int. Conf. on Image Analysis and Processing, Florence, Italy. (1997)
4. Iwase, S., Saito, H.: Tracking soccer player using multiple views. In: IAPR Workshop on Machine Vision Applications. (2002)
5. OK, H., Seo, Y., Hong, K.: Multiple soccer players tracking by condensation with occlusion alarm probability. In: Int. Workshop on Statistically Motivated Vision Processing, in conjunction with ECCV 2002, Copenhagen, Denmark. (2002)
6. Yoon, H., Bae, Y., Yang, Y.: A soccer image mosaicking and analysis method using line and advertisement board detection. ETRI Journal **24** (2002)
7. Utsumi, O., Miura, K., IDE, I., Sakai, S., Tanaka, H.: An object detection method for describing soccer games from video. In: IEEE International Conference on Multimedia and Expo (ICME). (2002)
8. Kang, J., Cohen, I., Medioni, G.: Soccer player tracking across uncalibrated camera streams. In: Joint IEEE International Workshop on Visual Surveillance and Performance Evaluation of Tracking and Surveillance (VS-PETS). (2003)
9. Yamada, A., Shirai, Y., Miura, J.: Tracking players and a ball in video image sequence and estimating camera parameters for 3d interpretation of soccer games. In: Proc. International Conference on Pattern Recognition. (2002)
10. Kitagawa, G.: Monte-carlo filter and smoother for non-gaussian nonlinear state space model. Journal of Computational and Graphical Statistics (1996)
11. Blake, A., Isard, M.: Active Contours. Springer-Verlag (1997)
12. Liu, J., Chen, R.: Sequential Monte Carlo methods for dynamic systems. (1998)
13. Doucet, A., Godsill, S., Andrieu, C.: On sequential monte-carlo sampling methods for bayesian filtering. (2000)
14. Doucet, A., Freitas, N.D., Gordon, N., eds.: Sequential Monte Carlo Methods in Practice. Springer-Verlag (2001)
15. McKenna, S.J., Raja, Y., Gong, S.: Object tracking using adaptive color mixture models. In: Proc. Asian Conf. on Computer Vision. (1998) 615–622
16. Bevilacqua, A.: A novel background initialization method in visual surveillance. In: IAPR Workshop on Machine Vision Applications. (2002)
17. Elgammal, A., Duraiswami, R., Harwood, D., Davis, L.S.: Background and foreground modeling using nonparametric kernel density for visual surveillance. In: Proceedings of the IEEE. Volume 90. (2002) 1151–1163
18. Lee, L.: Similarity-Based Approaches to Natural Language Processing. PhD thesis, Harvard University, Cambridge, MA (1997)

Multi-model Component-Based Tracking
Using Robust Information Fusion

Bogdan Georgescu[1], Dorin Comaniciu[1], Tony X. Han[2], and Xiang Sean Zhou[1]

[1]Integrated Data Systems Department, Siemens Corporate Research,
755 College Road East, Princeton, NJ 08540, USA
{bogdan.georgescu, dorin.comaniciu, xiang.zhou}@scr.siemens.com
[2]Beckman Institute and ECE Department, University of Illinois at Urbana-Champaign,
405 N. Mathews Ave., Urbana, IL 61801, USA
mailto:xuhan@ifp.uiuc.eduxuhan@ifp.uiuc.edu

Abstract. One of the most difficult aspects of visual object tracking is the handling of occlusions and target appearance changes due to variations in illumination and viewing direction. To address these challenges we introduce a novel tracking technique that relies on component-based target representations and on robust fusion to integrate model information across frames. More specifically, we maintain a set of component-based models of the target, acquired at different time instances, and combine robustly the estimated motion suggested by each component to determine the next position of the target. In this paper we allow the target to undergo similarity transformations, although the framework is general enough to be applied to more complex ones. We pay particular attention to uncertainty handling and propagation, for component motion estimation, robust fusion across time and estimation of the similarity transform. The theory is tested on very difficult real tracking scenarios with promising results.

1 Introduction

One of the problems of visual tracking of objects is to maintain a representation of target appearance that has to be robust enough to cope with inherent changes due to target movement and/or camera movement. Methods based on template matching have to adapt the model template in order to successfully track the target. Without adaptation, tracking is reliable only over short periods of time when the appearance does not change significantly. However, in most applications, for long time periods the target appearance undergoes considerable changes in structure due to change of viewpoint, illumination or it can be occluded. Methods based on motion tracking [1], [2], where the model is adapted to the previous frame, can deal with such appearance changes. However accumulated motion error and rapid visual changes make the model to drift away from the tracked target. Tracking performance can be improved by imposing object specific subspace constraints [3], [4] or maintaining a statistical representation of the model [5], [6], [7]. This representation can be determined a priori or computed on line. The appearance variability can be modeled as a probability distribution function which ideally is learned on line. Previous work approximated this p.d.f. as a normal

D. Comaniciu et al. (Eds.): SMVP 2004, LNCS 3247, pp. 61–70, 2004.

distribution in which case the mean represent the most likely model template. Updating the distribution parameters can be done using EM based algorithms. Also adaptive mixture models have been proposed to cope with outliers and sudden appearance changes [5].

We propose a method where the appearance variability is simply modeled by maintaining several models over time. This amounts for a nonparametric representation of the probability density function that characterizes the object appearance. We also adopt a component based approach and divide the target into several regions which are processed separately. Tracking is performed by obtaining independently from each model a motion estimate and its uncertainty through optical flow. A recently proposed robust fusion technique [8] is used to compute the final estimate for each component. The method, named Variable-Bandwidth Density-based Fusion (VBDF), computes the location of the most significant mode of the displacements density function while taking into account their uncertainty. The VBDF method manages the multiple data sources and outliers in the motion estimates. In this framework, occlusions are naturally handled through the estimate uncertainty for large residual errors. The alignment error is used to compute the scale of the covariance matrix of the estimate, therefore reducing the influence of the unreliable displacements.

The paper is organized as follows. Section 2 contains previous work on appearance modeling related to our approach. The multi-model component based tracking method is presented in Section 3. Experiments on real sequences under considerable occlusions are in Section 4 and we conclude in Section 5.

2   Related Work

An intrinsic characteristic of the vision based tracking is that the appearance of the tracking target and the background are inevitably changing, albeit gradually. Since the general invariant features for robust tracking are hard to find, most of the current methods need to handle the appearance variation of the tracking target and/or background. Every tracking scheme involve a certain representation of the 2D image appearance of the object, even though this is not mentioned explicitly.

Fleet et al. [5] proposed a generative model containing 3 components: the stable component, the wandering component, and the occlusion component. The stable component identifying the most reliable structure for motion estimation and the wandering component representing the variation of the appearance are two Gaussian distributions. The occlusion component accounting for data outliers is uniformly distributed on the possible intensity level. Using the phase parts of the steerable wavelet coefficients [9] as feature, this algorithm achieves satisfactory tracking results. It needs a relative long time for the stable component to gain confidence in appearance estimation. Since the stable component, as the tracking template, is modeled as an unimodal Gaussian, it needs to restart from time to time to accommodate the natural multimodal case.

Based on the hypothesis that most promising features for tracking are the same features that best discriminate between object and background classes, Collins et al. [10] empirically evaluate all candidate features to estimate the distributions and a new image composed of the log likelihood ratio of these distributions is used to track. However some salient feature that distinguish the tracking object from background may change drastically which imperils the validity of the hypothesis used in [10].

Using Earth Movers Distance as feature, Sharma et al. [11] present a complete object appearance learning approach, dealing with 3D surfaces. The problem is modeled by a continuous Markov Random Field with clique potentials defined as energy function. This approach safely maps and maintains object appearance on the 3D model surface. An online adapted appearance model is proposed in [12] using a Markov Random Field of the color distributions over a 3D model. Appearance driven cue confidences are used to balance the contribution for model update.

3 Multi-model Component-Based Tracker

Object tracking challenges due to occlusions and appearance variations are handled in our framework through a multi-model component-based approach. Maintaining several representatives for the 2D appearance model does not restrict it to a unimodal distribution and the VBDF fusion mechanism robustly integrates multiple estimates to determine the most dominant motion for each component. These key ideas are introduced in Subsection 3.1 followed by the VBDF algorithm in Subsection 3.2. Details about our method and implementation issues are in Subsection 3.3.

3.1 Main Ideas

The steps of our proposed method are outlined in Figure 1. To model the changes during tracking we propose to maintain several exemplars of the object appearance over time. This is in contrast to the approach adopted in [5] where the stable appearance component is modeled using a Gaussian distribution for each pixel. Maintaining explicitly the intensities is equivalent to a nonparametric representation of the appearance distribution.

The top row in Figure 1 illustrates the current exemplars in the model set, each having associated a set of overlapping components. A component-based approach is more robust that a global representation, being less sensitive to illumination changes and pose. Another advantage is that partial occlusion can be handled at the component level by analyzing the matching likelihood.

Each component is processed independently, its location and covariance matrix is estimated in the current image with respect to all of the model templates. For example, one of the components is illustrated by the gray rectangle in Figure 1 and its location and uncertainty with respect to each model is shown in I_{new}. The VBDF robust fusion procedure is applied to determine the most dominant motion (mode) with the associated uncertainty (Figure 1, center bottom row). Note the variance in the estimated location of each component due to occlusion or appearance change.

The location of the components in the current frame is further constrained by a global parametric motion model. We assume a similarity transformation model and its parameters are estimated using the confidence in each component location. Therefore the reliable components contribute more to the global motion estimation.

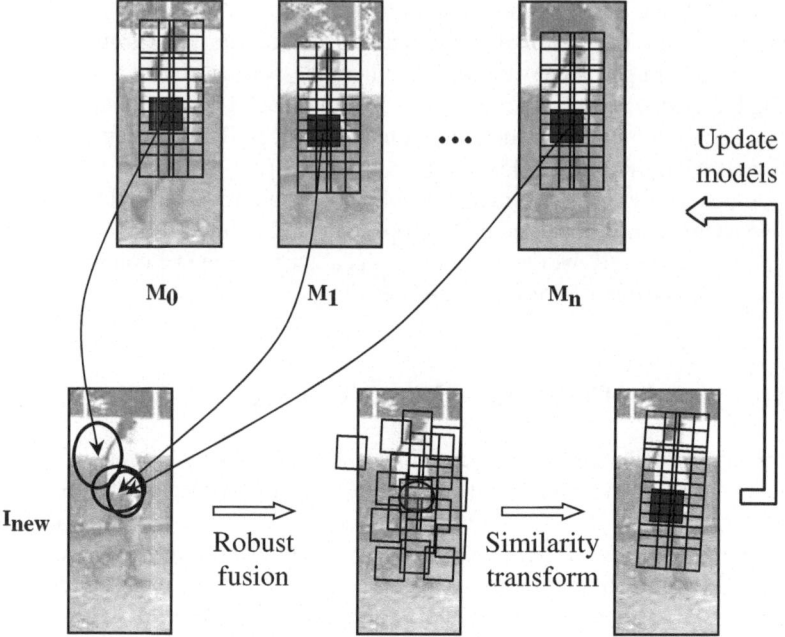

Fig. 1. The steps of the multi-model component based tracker

The current frame is added to the model set if the residual error to the reference appearances is relatively low. The threshold is chosen such that we do not add the images where the target has significant occlusion. The number of exemplars in our model set is fixed, therefore the oldest one is discarded.

3.2   Variable-Bandwidth Density-Based Fusion

The VDBF estimator is based on nonparametric density estimation with adaptive kernel bandwidths. It was introduced in [8] with an application to robust optical flow computation. The choice of the VDBF estimator is motivated by its good performance in the presence of outliers in the input data when compared to previously proposed methods such as *Covariance Intersection* [13] or BLUE estimate assuming single source, statistically independent data [14]. The robustness with respect to outliers of the VDBF technique comes from the nonparametric estimation of the initial data distribution while exploiting its uncertainty. The VBDF estimator is defined as the location of the most significant mode of the density function. The mode computation is based on the variable-bandwidth mean shift technique in a multiscale optimization framework.

Let $x_i \in \mathbb{R}^d$, $i = 1 \ldots n$ be the available d-dimensional estimates, each having an associated uncertainty given by the covariance matrix C_i. The most significant mode of the their density function is determined iteratively in a multiscale fashion. A bandwidth matrix $H_i = C_i + \alpha^2 I$ is associated with each point x_i, where I is the identity matrix and the parameter α determines the scale of the analysis. The sample point density estimator at location x is defined by

$$\hat{f}(\boldsymbol{x}) = \frac{1}{n(2\pi)^{d/2}} \sum_{i=1}^{n} exp\left(-\frac{1}{2}D^2(\boldsymbol{x}, \boldsymbol{x}_i, \mathrm{H}_i)\right) \qquad (1)$$

where D represents the Mahalanobis distance between $\boldsymbol{x}$ and $\boldsymbol{x}_i$

$$D^2(\boldsymbol{x}, \boldsymbol{x}_i, \mathrm{H}_i) = (\boldsymbol{x} - \boldsymbol{x}_i)^\top \mathrm{H}_i^{-1}(\boldsymbol{x} - \boldsymbol{x}_i) \qquad (2)$$

The variable bandwidth mean shift vector at location $\boldsymbol{x}$ is given by

$$\boldsymbol{m}(\boldsymbol{x}) = \mathrm{H}_h(\boldsymbol{x}) \sum_{i=1}^{n} \omega_i(\boldsymbol{x})\mathrm{H}_i^{-1}\boldsymbol{x}_i - \boldsymbol{x} \qquad (3)$$

where H_h represents the harmonic mean of the bandwidth matrices weighted by the data-dependent weights $\omega_i(\boldsymbol{x})$

$$\mathrm{H}_h(\boldsymbol{x}) = \left(\sum_{i=1}^{n} \omega_i(\boldsymbol{x})\mathrm{H}_i^{-1}\right)^{-1}. \qquad (4)$$

The data dependent weights computed at the current location $\boldsymbol{x}$ have the expression

$$\omega_i(\boldsymbol{x}) = \frac{\frac{1}{|\mathrm{H}_i|^{1/2}}exp\left(-\frac{1}{2}D^2(\boldsymbol{x}, \boldsymbol{x}_i, \mathrm{H}_i)\right)}{\sum_{i=1}^{n} \frac{1}{|\mathrm{H}_i|^{1/2}}exp\left(-\frac{1}{2}D^2(\boldsymbol{x}, \boldsymbol{x}_i, \mathrm{H}_i)\right)} \qquad (5)$$

and note that they satisfy $\sum_{i=1}^{n} \omega_i(\boldsymbol{x}) = 1$.

It can be shown that the density corresponding to the point $\boldsymbol{x} + \boldsymbol{m}(\boldsymbol{x})$ is always higher or equal to the one corresponding to $\boldsymbol{x}$. Therefore iteratively updating the current location using the mean shift vector yields a hill-climbing procedure which converges to a stationary point of the underlying density.

The VBDF estimator finds the most important mode by iteratively applying the adaptive mean shift procedure at several scales. It starts from a large scale by choosing the parameter α large with respect to the spread of the points $\boldsymbol{x}_i$. In this case the density surface is unimodal therefore the determined mode will correspond to the globally densest region. The procedure is repeated while reducing the value of the parameter α and starting the the mean shift iterations from the mode determined at the previous scale. For the final step the bandwidth matrix associated to each point is equal to the covariance matrix, i.e. $\mathrm{H}_i = \mathrm{C}_i$.

The VBDF estimator is a powerful tool for information fusion with the ability to deal with multiple source models. This is important for motion estimation as points in a local neighborhood may exhibit multiple motions. The most significant mode corresponds to the most relevant motion.

3.3 Tracking Multiple Component Models

Consider that we have n models $\mathrm{M}_0, \mathrm{M}_1, \ldots, \mathrm{M}_n$. For each image we maintain c components with their location denoted by $\boldsymbol{x}_{ij}$, $i = 1 \ldots n$, $j = 1 \ldots c$. When a new image is available we estimate the location and the uncertainty for each component and for each model. This step can be done using several techniques such as ones based on image correlation, spatial gradient or regularization of spatio-temporal energy. Based on the image brightness constancy, one of the most popular optical flow techniques has been developed by Lucas and Kanade [15]. For a small image patch the pixels flow estimates are combined assuming a translational model by solving a weighted least squares problem. However they neglect the uncertainty of the initial estimates, therefore we adopt the robust optical flow technique proposed in [8] which is also an application of the VBDF technique. The result is the motion estimate $\hat{x}_{ij}$ for each component and its uncertainty $\hat{C}_{ij}$. Thus $\hat{x}_{ij}$ represents the location estimate of component j with respect to model i. The scale of the covariance matrix is also estimated from the matching residual errors. This will increase the size of the covariance matrix when the respective component is occluded therefore we naturally handle occlusions at the component level.

The VBDF robust fusion technique presented in the previous subsection is applied to determine the most relevant location $\hat{x}_j$ for component j in the current frame. The mode tracking across scales results in

$$\hat{x}_j = C(\hat{x}_j) \sum_{i=1}^{n} \omega_i(\hat{x}_j) \hat{C}_{ij}^{-1} \hat{x}_{ij}$$

$$C(\hat{x}_j) = \left(\sum_{i=1}^{n} \omega_i(\hat{x}_j) \hat{C}_{ij}^{-1} \right)^{-1} . \tag{6}$$

with the weights ω_i defined as in (5).

Following the location computation of each component, a weighted rectangle fitting is carried out with the weights given by the covariance matrix of the estimates. We assume that the image patches are related by a similarity transform T defined by 4 parameters. The similarity transform of the dynamic component location x is characterized by the following equations.

$$T(x) = \begin{pmatrix} a & -b \\ b & a \end{pmatrix} x + \begin{pmatrix} t_x \\ t_y \end{pmatrix} \tag{7}$$

where t_x, t_y are the translational parameters and a, b parametrize the 2D rotation and scaling.

The minimized criterion is the sum of Mahalanobis distances between the reference location x_j^0 and the estimated ones $\hat{x}_j$ (j^{th} component location in the current frame).

$$\mathcal{J} = \sum_{j=1}^{c} (\hat{x}_j - T(x_j^0))^T C(\hat{x}_j)^{-1} (\hat{x}_j - T(x_j^0)) . \tag{8}$$

Minimization is done through standard weighted least squares. Note that because we use the covariance matrix for each component the influence of points with high uncertainty is reduced.

After the rectangle is fitted to the tracked components, we uniformly resample the dynamic component candidate inside the rectangle. We assume the relative position of each component with respect to the rectangle does not change a lot. If the distance of the resample position and the track position computed by the optical flow of a certain component is larger than a tolerable threshold, we regard the tracked position as an outlier and replace it with the resampled point. The current image is added to the model set if sufficient components have low residual error. The median residual error between the models and the current frame is compared with a pre-determined threshold T_h.

Given a set of models $M_0, M_1, \ldots, M_n$ in which the component j has location x_{ij} in frame i, our object tracking algorithm can be summarized by the following steps:

1. Given a new image I_f compute $\hat{x}_{ij}^{(f)}$ through robust optical flow [8] starting from $\hat{x}_j^{(f-1)}$, the location estimated in the previous frame;
2. For $j = 1 \ldots c$ estimate the location $\hat{x}_j^{(f)}$ of component j using the VBDF estimator (Subsection 3.2) resulting in (6);
3. Constrain the component location using the transform computed by minimizing (8);
4. Add the new appearance to the model set if its median residual error is less that T_h.

The proposed multi-template framework can be directly applied in the context of shape tracking. If the tracked points represent the control points of a shape modeled by splines, the use of the robust fusion of multiple position estimates increases the reliability of the location estimate of the shape. It also results in smaller corrections when the shape space is limited by learned subspace constraints. If the contour is available, the model templates used for tracking can be selected online from the model set based on the distance between shapes.

4   Experiments

The proposed method was applied for object tracking in real videos with significant clutter and occlusions. We used $n = 20$ model templates and the components are at 5 pixels distance with their number c determined by the bounding rectangle. The threshold T_h for a new image to be added to the model set was $1/8$ of the intensity range. The value was learned from the data such that occlusions are detected.

We successfully tested the method on different image sequences including medical videos. We present results on only two sequences that illustrate the advantages of our approach.

The results for tracking a person's face is presented in Figure 2. This is a very challenging sequence where the scene has significant clutter with several faces and multiple occlusions affect the tracked region. Figure 3 shows the median residual error over time which is used for model update. The peaks in the graph correspond to frames where the target is completely occluded. As mentioned earlier, the model update occurs when the error passes the threshold $T_h = 32$ which is the horizontal line in Figure 3.

Figure 4 shows the results of tracking of a human body. The method is able to cope with the appearance changes such as the arm moving and it is able to recover the tracking target after the tree occlusion. Figure 5 plots the residual error over time. The first peak corresponds to the target being occluded by the tree while toward the end the

Fig. 2. Face tracking results; the white rectangle represents the target. (a) Frame 0; (b) Frame 49; (c) Frame 137; (d) Frame 249; (e) Frame 281; (f) Frame 312; (g) Frame 375; (h) Frame 527

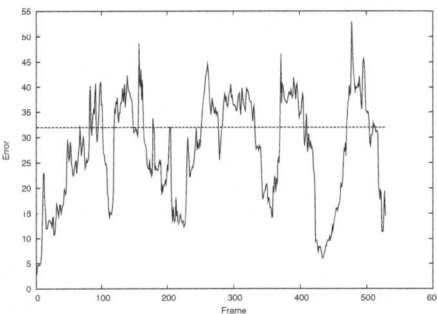

Fig. 3. Residual error over time for face tracking sequence.Horizontal line represents the model update threshold

Fig. 4. Human body tracking results.(a) Frame 0; (b) Frame 27; (c) Frame 38; (d) Frame 106

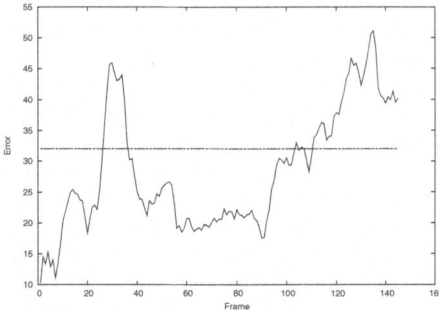

Fig. 5. Residual error over time for human body tracking sequence. Horizontal line represents the model update threshold

error is due to the person turning and its image size becoming smaller with respect to the fixed component size.

5   Conclusion

This paper introduced an object tracking method based on multiple appearance models and VBDF-based fusion of estimates. We showed the ability of the proposed approach to deal with significant occlusions, clutter and appearance changes on real image sequences. Although we used image templates as models, our approach is general enough to integrate information from different model representations such as color distributions or filter responses. Further work include solving for the global motion through a robust approach and use of multiple hypothesis for tracking.

References

1. Shi, J., Tomasi, C.: Good features to track. In: 1994 IEEE Conf. on Computer Vision and Pattern Recog., San Juan, Puerto Rico (1994) 593–600
2. Sidenbladh, H., Black, M.J., Fleet, D.J.: Stochastic tracking of 3d human figures using 2d image motion. In: 2000 European Conf. on Computer Vision. Volume 2., Dublin, Ireland (2000) 702–718
3. Black, M.J., Jepson, A.D.: Eigentracking: Robust matching and tracking of articulated objects using a view-based representation. International J. of Computer Vision 26 (1998) 63–84
4. Edwards, G.J., Cootes, T.F., Taylor, C.J.: Face recognition using active appearance models. In: 1998 European Conf. on Computer Vision, Freiburg, Germany (1998) 581–595
5. Jepson, A.D., Fleet, D.J., El-Maraghi, T.F.: Robust online appearance models for visual tracking. IEEE Trans. Pattern Anal. Machine Intell. 25 (2003) 1296–1311
6. Stauffer, C., Grimson, W.E.L.: Adaptive background mixture models for real-time tracking. In: 1999 IEEE Conf. on Computer Vision and Pattern Recog. Volume 2. (1999) 246–252
7. Tao, H., Sawhney, H.S., Kumar, R.: Dynamic layer representation with application to tracking. In: 2000 IEEE Conf. on Computer Vision and Pattern Recog. Volume 2. (2000) 134–141
8. Comaniciu, D.: Nonparametric information fusion for motion estimation. In: 2003 IEEE Conf. on Computer Vision and Pattern Recog. Volume I., Madison, WI (2003) 59–66
9. Freeman, W.T., Adelson, E.H.: The design and use of steerable filters. IEEE Trans. Pattern Anal. Machine Intell. 13 (1991) 891–906
10. Collins, R., Liu, Y.: On-line selection of discriminative tracking features. In: 2003 International Conf. on Computer Vision. (2003)
11. Krahnstoever, N., Sharma, R.: Robust probabilistic estimation of uncertain appearance for model based tracking. In: IEEE Workshop on Motion and Video Computing. (2002)
12. Krahnstoever, N., Sharma, R.: Appearance management and cue fusion for 3d model-based tracking. In: 2003 IEEE Conf. on Computer Vision and Pattern Recog., Madison, WI (2003)
13. Julier, S., Uhlmann, J.: A non-divergent extimation algorithm in the presence of unknown correlations. In: Proc. American Control Conf., Alberqueque, NM (1997)
14. Singh, A., Allen, P.: Image-flow computation: An estimation-theoretic framework and a unified perspective. CVGIP: Image Understanding 56 (1992) 152–177
15. Lucas, B., Kanade, T.: An iterative image registration technique with application to stereo vision. In: International Joint Conf. on Artificial Intelligence, Vancouver, Canada (1981) 674–679

A Probabilistic Approach to Large Displacement Optical Flow and Occlusion Detection

Christoph Strecha[1], Rik Fransens[1], and Luc Van Gool[1]

K.U.Leuven, Belgium
firstname.surname@esat.kuleuven.ac.be
http://www.esat.kuleuven.ac.be/psi/visics

Abstract. This paper deals with the computation of optical flow and occlusion detection in the case of large displacements. We propose a Bayesian approach to the optical flow problem and solve it by means of differential techniques. The images are regarded as noisy measurements of an underlying 'true' image-function. Additionally, the image data is considered incomplete, in the sense that we do not know which pixels from a particular image are occluded in the other images. We describe an EM-algorithm, which iterates between estimating values for all hidden quantities, and optimizing the current optical flow estimates by differential techniques. The Bayesian way of describing the problem leads to more insight in existing differential approaches, and offers some natural extensions to them. The resulting system involves less parameters and gives an interpretation to the remaining ones. An important new feature is the photometric detection of occluded pixels. We compare the algorithm with existing optical flow methods on ground truth data. The comparison shows that our algorithm generates the most accurate optical flow estimates. We further illustrate the approach with some challenging real-world examples.

1 Introduction

A fundamental problem in the processing of image sequences is the computation of optical flow. Optical flow is caused by the time-varying projection of objects onto a possibly moving image plane. It is therefore the most general transformation, assigning a two dimensional displacement vector to every pixel. There exist other image motion representations that restrict optical flow to parametric models (affine, homography) or to one dimensional disparities as in the case of stereo. Optical flow estimation has been extensively studied. The most common methods are differential, frequency-based, block matching and correlation-based methods. We refer the interested reader to Barron *et al.* [4] for a description and comparison of important optical flow techniques.

In this paper we focus on differential techniques and investigate the explicit detection of occluded pixels. Differential techniques were, in the context of optical flow computation, introduced by Horn and Schunck [7]. They are based on the Optical Flow Constraint (OFC) equation that relates spatial and temporal brightness gradients to the two components of the optical flow field. Additional constraints on this displacement field have to be added to overcome the ill-possedness of the problem. Originally Horn

D. Comaniciu et al. (Eds.): SMVP 2004, LNCS 3247, pp. 71–82, 2004.

and Schunck used an isotropic smoothness constraint, which is not able to deal with discontinuities. Other researchers investigated anisotropic versions of smoothness based on image gradients [9, 2] and optical flow gradients [5, 3]. In these approaches, new parameters controlling the strength of anisotropy are introduced. A taxonomy of different smoothness constraints, of which some are also used in nonlinear diffusion filtering, is presented in [12]. Proesmans *et al.*[10] present a third way to introduce anisotropy. They define a consistency function that depends on the difference of matching image one to image two and vice verse (so-called *forward-backward* matching). In their work, occluded pixels are characterized by a low value of the consistency function. A similar forward-backward strategy for occlusion detection, but with structure tensor regularization, is used by Alvarez *et al.* [1]. However, these authors require the forward-backward optical flow computation *and* additional parameters for solving the occlusion problem.

We formulate the correspondence problem in a probabilistic framework. This results in an EM algorithm, whose maximization step involves diffusion equations similar to the ones previously described. However, the probabilistic description now gives an interpretation to the most important parameter (λ) which controls the balancing between image matching and flow field smoothness. The formulation leads naturally to the detection of occlusions based on the image values themself, which prevents the computation of the two (forward-backward) optical flow fields. Our algorithm needs no additional parameters for detecting occlusions, and the relative contribution of the spectral components to the matching term is handled automatically.

This paper is organized as follows. We proceed with the explanation of the probabilistic framework and its EM-based solution. Section 3 deals with the actual optical flow computation. We introduce two algorithms that differ by their anisotropic smoothness term and discuss the implications of the probabilistic view to the matching term. Experiments on ground truth and real data are discussed in section 4, where we compare our two algorithms with an implementation of Alvarez *et al.*[2].

2   Probabilistic Account for Optical Flow and Occlusions Estimation

Suppose we are given two images $\mathcal{I}_1$ and $\mathcal{I}_2$, which associate a 2D-coordinate $\mathbf{x}$ with an image value $\mathcal{I}_i(\mathbf{x})$. If we are dealing with color images, this value is a 3-vector and for intensity images it is a scalar[1]. Our goal is to estimate the two-dimensional displacement or optical flow field $\mathcal{F}(\mathbf{x})$ such that $\mathcal{I}_1(\mathbf{x})$ is brought into correspondence with $\mathcal{I}_2(\mathbf{x} + \mathcal{F}(\mathbf{x}))$.

Faithful to the Bayesian philosophy, we regard the input images as noisy measurements of an unknown image irradiance $\mathcal{I}_1^*$. This allows us to write:

$$\mathcal{I}_1(\mathbf{x}) = \mathcal{I}_1^*(\mathbf{x}) + \epsilon \tag{1}$$

$$\mathcal{I}_2(\mathbf{x} + \mathcal{F}(\mathbf{x})) = \mathcal{I}_1^*(\mathbf{x}) + \epsilon \tag{2}$$

$$\epsilon \sim \mathcal{N}(\mathbf{0}, \mathbf{\Sigma}) \tag{3}$$

---

[1] In fact one could add other features such as filter responses to the image [5]. We continue the discussion for general n-band images.

where $\mathcal{N}()$ is a normal noise distribution with zero-mean and covariance matrix Σ, which we assume to be equal for all images. Both the irradiance or 'true' image $\mathcal{I}_1^*$ and Σ are unknown, and their estimation becomes part of the optimization procedure.

A major complication for larger displacements is the occlusion problem, which arises from the fact that not all parts of the scene, which are visible in a particular image, are also visible in the other images due to occlusion. When computing image correspondences, such occluded regions should be identified and excluded from the matching procedure. This will be modelled by introducing a visibility map $\mathcal{V}(\mathbf{x})$, which signal whether a scene point X that projects onto $\mathbf{x}$ in $\mathcal{I}_1$ is also visible in image $\mathcal{I}_2$ or not. Every element of $\mathcal{V}(\mathbf{x})$ is a binary random variable which is either 1 or 0, corresponding to visibility or occlusion, respectively. $\mathcal{V}(\mathbf{x})$ is a hidden variable, and its value must be inferred from the input images.

Estimating the optical flow field $\mathcal{F}(\mathbf{x})$ can now be formally stated as finding those flow values which make the image correspondences $\mathcal{I}_1^*(\mathbf{x}) \Leftrightarrow \mathcal{I}_2(\mathbf{x} + \mathcal{F}(\mathbf{x}))$ restricted to $\mathcal{V}(\mathbf{x}) = 1$, most probable.

2.1 MAP Estimation

We are now facing the hard problem of estimating the unknown quantities $\theta = \mathcal{F}, \mathcal{I}_1^*$ and Σ given the images $\mathcal{I}_1$ and $\mathcal{I}_2$. Furthermore, we have introduced the unobservable or hidden variables $\mathcal{V}$, which must also be inferred over the course of the optimization. In a Bayesian framework, the optimal value for θ is the one that maximizes the posterior probability $p(\theta|\mathcal{I}_1, \mathcal{I}_2)$. According to Bayes' rule, this posterior can be written as:

$$p(\theta|\mathcal{I}_1, \mathcal{I}_2) = \frac{\int p(\mathcal{I}_1, \mathcal{I}_2 | \theta, \mathcal{V}) p(\theta | \mathcal{V}) p(\mathcal{V}) d\mathcal{V}}{p(\mathcal{I}_1, \mathcal{I}_2)} , \qquad (4)$$

where we have conditioned the data likelihood and the prior on the hidden variables $\mathcal{V}$. The denominator or 'evidence' is merely the integral of the numerator over all possible values of θ and can be ignored in the maximization problem. Hence, we will try to optimize the numerator only. In order to find the most probable value for θ, we need to integrate over all possible values of $\mathcal{V}$ which is computationally intractable. Instead, we assume that the probability density function (PDF) of $\mathcal{V}$ is peaked about a single value, i.e. $p(\mathcal{V})$ is a Dirac-function centered at this value. This leads to an Estimation-Maximization (EM) based solution, which iterates between (i) estimating values for $\mathcal{V}$, given the current estimate of θ, and (ii) maximizing the posterior probability of θ, given the current estimate of $\mathcal{V}$. A more detailed description of this procedure will be given later. So, given a current estimate $\hat{\mathcal{V}}$ for the hidden variables, we want to optimize:

$$q(\theta|\mathcal{I}_1, \mathcal{I}_2) = p(\mathcal{I}_1, \mathcal{I}_2 | \theta, \hat{\mathcal{V}}) p(\theta | \hat{\mathcal{V}}) \qquad (5)$$

The a-posteriori probability of θ is proportional to the product of two terms: the data-likelihood $p(\mathcal{I}_1, \mathcal{I}_2 | \theta, \hat{\mathcal{V}})$ and a prior $p(\theta|\hat{\mathcal{V}})$, which we call L and P, respectively. We now discuss both terms in turn.

Under the assumption that the image noise is i.i.d. for all pixels in both views, the data likelihood L can be written as the product of all individual pixel probabilities:

$$L = \prod_{\mathbf{x}} p(\mathcal{I}_1(\mathbf{x})|\theta) \prod_{\mathbf{x}} p(\mathcal{I}_2(\mathbf{x} + \mathcal{F}(\mathbf{x}))|\theta) , \qquad (6)$$

where the second product is restricted to those $\mathbf{x}$ for which $\mathcal{V}(\mathbf{x}) = 1$. Note that, by definition, all pixels in $\mathcal{I}_1^*$ are visible in $\mathcal{I}_1$. Given the current estimate of the 'true' image $\mathcal{I}_1^*(\mathbf{x})$ and the noise distribution $\boldsymbol{\Sigma}$, we can further specify the likelihood by the normal distribution $\mathcal{N}$:

$$L = \prod_{i=1}^{2} \prod_{\mathbf{x}} \frac{1}{(2\pi)^{d/2}|\boldsymbol{\Sigma}|^{1/2}} \exp\left(-\frac{1}{2}\mathbf{m}_i(\mathbf{x})^T \boldsymbol{\Sigma}^{-1} \mathbf{m}_i(\mathbf{x})\right), \tag{7}$$

where $\mathbf{m}_1(\mathbf{x}) = \mathcal{I}_1^*(\mathbf{x}) - \mathcal{I}_1(\mathbf{x})$ and $\mathbf{m}_2(\mathbf{x}) = \mathcal{I}_1^*(\mathbf{x}) - \mathcal{I}_2(\mathbf{x}+\mathcal{F}(\mathbf{x}))$ are the differences of the true image function with the input images $\mathcal{I}_1$ and $\mathcal{I}_2$, the latter one estimated at the current value of $\mathcal{F}$. The variable d in the normalization constant denotes the dimensionality of $\mathbf{m}_i$.

The formulation of an appropriate prior is slightly more complicated. We can marginalize P as the product of a displacement $\mathcal{F}$ dependent and image dependent part:

$$P = p(\mathcal{F}|\mathcal{I}_1^*, \boldsymbol{\Sigma})p(\mathcal{I}_1^*, \boldsymbol{\Sigma}). \tag{8}$$

Assuming we have no prior preference for the image related parameters, i.e. assuming a uniform prior over $\mathcal{I}_1^*$ and $\boldsymbol{\Sigma}$, this can be rewritten as:

$$P = p(\mathcal{F}|\mathcal{I}_1^*, \boldsymbol{\Sigma})c \tag{9}$$

where c is an appropriate constant.

The displacement prior $p(\mathcal{F} \mid \mathcal{I}_1^*, \boldsymbol{\Sigma})$ will be modelled as an exponential density distribution of the form $exp(-R(\mathcal{I}_1^*, \mathcal{F})/\lambda)$. Here, λ is a parameter which controls the width of the distribution, and $R(\mathcal{I}_1^*, \mathcal{F})$ is a data-driven 'regularizer'. From such a regularizer we expect that it reflects our prior belief that the world is essentially simple, i.e. for a locally smooth solution $\mathcal{F}$ in the neighborhood of a particular point $\mathbf{x}$, its value should approach zero, making such a solution very likely. Vice-versa, large fluctuations of the optical flow field should result in large values for the regularizer, making such solutions less likely. Furthermore, the regularizer should be data-driven: if the image $\mathcal{I}_1^*$ suggests a discontinuity, i.e. by the presence of a high image gradient at a particular point $\mathbf{x}$, a large discontinuity at $\mathbf{x}$ should not be made a-priori unlikely. Such regularizers are commonly used in the PDE-community [2, 9], where they serve as *image driven anisotropic diffusion operators* in optical flow or edge-preserving smoothing computations. The regularizer is given by:

$$R(\mathcal{I}_1^*, \mathcal{F}) = \nabla \mathcal{F}^T T(\nabla \mathcal{I}_1^*) \nabla \mathcal{F}. \tag{10}$$

Here, $T(\nabla \mathcal{I}_1^*)$ is a diffusion tensor defined by:

$$T(\nabla \mathcal{I}_1^*) = \frac{1}{|\nabla \mathcal{I}_1^*|^2 + 2\nu^2}\left(\nabla \mathcal{I}_1^{*\perp} \nabla \mathcal{I}_1^{*\perp T} + \nu^2 \mathbf{1}\right), \tag{11}$$

where $\mathbf{1}$ is the identity matrix, ν is a parameter controlling the degree of anisotropy and $\nabla \mathcal{I}_1^{*\perp}$ is the vector perpendicular to $\nabla \mathcal{I}_1^*$. For color images, the tensor is defined as the sum of the 3 individual color channel tensors. $R(\mathcal{I}_1^*, \mathcal{F})$ is low when $\nabla \mathcal{F}$ is parallel to $\nabla \mathcal{I}_1^*$, which is exactly the desired behavior. Note that, by making the displacement

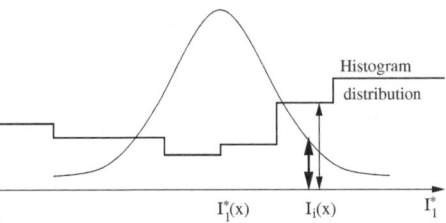

Fig. 1. Visibility estimation: The probability of $\mathcal{I}_i(\mathbf{x})$ being visible, is proportional to its value under the Gauss-curve, and the probability of $\mathcal{I}_i(\mathbf{x})$ being invisible is measured as the value under the histogram-based estimator

prior dependent on $\mathcal{I}_1^*$, it implicitly also makes it dependent on the original image data. Strictly speaking, this violates the Bayesian principle that priors should not be estimated from the data. In practice, however, it leads to more sensible solutions than setting them arbitrarily, or using so-called *conjugate* priors, whose main justification comes from computational simplicity [11].

We can now turn back to the optimization of $\boldsymbol{\theta}$. Instead of maximizing the posterior in (5), we minimize its negative logarithm. This leads (upto a constant) to the following energy:

$$E[\boldsymbol{\theta}] = \frac{1}{2} \sum_{i=1}^{2} \sum_{\mathbf{x}} \mathcal{V}(\mathbf{x}) \big[\mathbf{m}_i(\mathbf{x})^T \boldsymbol{\Sigma}^{-1} \mathbf{m}_i(\mathbf{x}) + \log((2\pi)^{\frac{d}{2}} |\boldsymbol{\Sigma}|) \big] + \frac{1}{\lambda} R(\mathcal{I}_1^*(\mathbf{x}), \mathcal{F}(\mathbf{x}))$$

$$(12)$$

2.2 An EM Solution

In the previous paragraph, an energy equation, w.r.t. the unknown quantities $\boldsymbol{\theta}$, was derived. This energy corresponds to the negative logarithm of the posterior distribution of $\boldsymbol{\theta}$, given the current estimate of the hidden variable $\mathcal{V}$. Now we will derive the EM-equations, which iterate between the estimation of $\mathcal{V}$ and the minimization of $E(\boldsymbol{\theta})$.

E-Step. On the $(k+1)^{th}$ iteration, the hidden variable $\mathcal{V}(\mathbf{x})$, are replaced by their conditional expectation given the data, where we use the current estimates $\boldsymbol{\theta}^{(k)}$ for $\boldsymbol{\theta}$. The expected value for the visibility is given by $E[\mathcal{V}|\mathcal{I}_1^*, \boldsymbol{\Sigma}, \mathcal{F}] \equiv \Pr(\mathcal{V}=1|\mathcal{I}_1^*, \boldsymbol{\Sigma}, \mathcal{F})$. According to Bayes' rule, the latter probability can be expressed as:

$$\Pr(\mathcal{V}=1|\mathcal{I}_1^*, \boldsymbol{\Sigma}, \mathcal{F}) = \frac{p(\mathcal{F}|\mathcal{V}=1, \mathcal{I}_1^*, \boldsymbol{\Sigma})}{p(\mathcal{F}|\mathcal{V}=1, \mathcal{I}_1^*, \boldsymbol{\Sigma}) + p(\mathcal{F}|\mathcal{V}=0, \mathcal{I}_1^*, \boldsymbol{\Sigma})} , \qquad (13)$$

where we have assumed equal priors on the probability of a pixel being visible or not. Given the current estimate of $\boldsymbol{\theta}$, the PDF $p(\mathcal{F}|\mathcal{V}=1, \mathcal{I}_1^*, \boldsymbol{\Sigma})$ is given by the value of the noise distribution evaluated over the color-difference between $\mathcal{I}_1^*(\mathbf{x})$ and $\mathcal{I}_2(\mathbf{x} + \mathcal{F}(\mathbf{x}))$:

$$p(\mathcal{F}|\mathcal{V}=1, \mathcal{I}_1^*, \boldsymbol{\Sigma}) = \mathcal{N}(\mathbf{m}_2; 0, \boldsymbol{\Sigma}) . \qquad (14)$$

The second PDF is more difficult to estimate, because it is hard to say what the color distribution of a pixel, which has no real counter-part in $\mathcal{I}_2$, looks like. We provide a

Table 1. Outline of the algorithm

1. Initialize $\mathcal{V} = 0.5$
2. Loop over pyramidal images
M-step until convergence
Compute $\mathcal{I}_1^*$ and Σ by eq.(16)
Compute $\mathcal{F}$ by solving the diffusion equation (see section 3)
E-Step
Estimate new visibilities $\mathcal{V}$ by eq. (15)
3. Rescale, goto next pyramidal level

global estimate for the PDF of occluded pixels by building a histogram of the color-values in $\mathcal{I}_1^*$ which are currently invisible. This is merely the histogram of $\mathcal{I}_1^*$ where the contribution of each pixel is weighted by $(1 - \mathcal{V}(\mathbf{x}))$. Note that, if a particular pixel in $\mathcal{I}_1^*$ is marked as not-visible, in the next iterations this will automatically decrease the visibility estimates of all similarly colored pixels. This makes sense from a perceptual point of view, and has a regularizing effect on the visibility maps. The update equations for $\mathcal{V}(\mathbf{x})$ are now:

$$\mathcal{V} \leftarrow \frac{\mathcal{N}(\mathbf{m}_2; \mathbf{0}, \Sigma)}{\mathcal{N} + \text{HIST}_{\mathcal{I}_1^*, (1-\mathcal{V})}(\mathcal{I}_1^*)} , \qquad (15)$$

where $\mathcal{N}$ is evaluated as in (14). This is graphically depicted in fig. (1).

M-Step. At the M-step, the intent is to compute values for $\boldsymbol{\theta}$ that maximizes (12), given the current estimates of $\mathcal{V}$. This is achieved by setting the parameters $\boldsymbol{\theta}$ to the appropriate root of the derivative equation, $\partial E(\boldsymbol{\theta})/\partial \theta = 0$.

For the image related parameters $\mathcal{I}_1^*$ and Σ, a closed form expressions for the roots can be derived and the update equations are:

$$\mathcal{I}_1^*(\mathbf{x}) \leftarrow \frac{1}{1 + \mathcal{V}(\mathbf{x})} \left(\mathcal{I}_1(\mathbf{x}) + \mathcal{V}(\mathbf{x})\mathcal{I}_2(\mathbf{x} + \mathcal{F}(\mathbf{x})) \right)$$

$$\Sigma \leftarrow \frac{1}{\sum_x (1 + \mathcal{V}(\mathbf{x}))} \sum_x \mathbf{m}_1(\mathbf{x})\mathbf{m}_1(\mathbf{x})^T + \mathcal{V}(\mathbf{x})\mathbf{m}_2(\mathbf{x})\mathbf{m}_2(\mathbf{x})^T . \qquad (16)$$

In order to arrive at these expressions, we ignored the effects of $\mathcal{I}_1^*$ and Σ on the regularization term. This is admissible because their influence on R is small compared to their influence on the matching term. Σ is only indirectly related to R through the computation of the visibility maps, which have an effect on R via the computation of $\mathcal{I}_1^*$. The image $\mathcal{I}_1^*$ has an effect on R via its gradient, which is used to define a quadratic norm on the depth gradient (10). Changes of $\mathcal{I}_1^*$ will therefore only have a minor influence on R.

However, for the update of the optical flow field $\mathcal{F}$ we are not so lucky, because $\mathcal{F}$ strongly influences both the matching and the regularization term. To minimize E w.r.t. $\mathcal{F}$, we solve the diffusion equation that can be derived from eq. 12 using the Euler-Lagrange formalism. The outline of the overall algorithm is given in table 1.

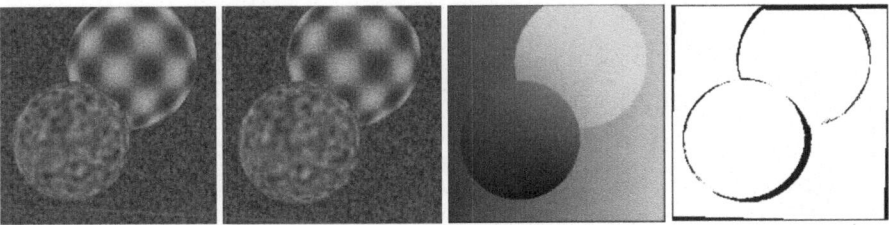

Fig. 2. One of the synthetic scenes used in the experiments. Left two images: input; middle right: ground truth optical flow; right: ground truth occlusions

3 Variational Optical Flow Estimation

In this section we will derive two update equations for the optical flow field. The first one, with image based anisotropic smoothness, follows directly from the previous considerations. The second one has a visibility based anisotropic smoothness term. We will also discuss the differences between these approaches and other differential optical flow techniques that have the ability to deal with large displacements.

3.1 Image Induced Anisotropy

Consider the main result of the first section - the energy in eq.(12) - where we only consider the terms depending on $\mathcal{F}(\mathbf{x})$:

$$E[\mathcal{F}(\mathbf{x})] = \sum_{\mathbf{x}} \mathcal{V}(\mathbf{x})\mathbf{m}_2(\mathbf{x})^T \mathbf{\Sigma}^{-1}\mathbf{m}_2(\mathbf{x}) + \frac{1}{\lambda}\nabla\mathcal{F}(\mathbf{x})^T\, T(\nabla\mathcal{I}_1^*)\, \nabla\mathcal{F}(\mathbf{x})\ . \quad (17)$$

The visibilities $\mathcal{V}$ are fixed for a particular instance of the M-step. For a given maximization step the three unknowns are estimated in turn while keeping the others fixed. In that case the minimum of the above energy is given by the Euler-Lagrange equation. To allow for large displacements, we split the value of $\mathcal{F}(\mathbf{x})$ into a current and a residual estimate [10, 2], i.e. $\mathcal{F}(\mathbf{x}) = \mathcal{F}_0(\mathbf{x}) + \mathcal{F}_r(\mathbf{x})$. Cutting of terms $\geq O(\mathcal{F}_r^2)$ from the Taylor expansion of $\mathbf{m}_2(\mathbf{x})$, we get:

$$\mathbf{m}_2(\mathbf{x}) = \mathcal{I}_1^* - \mathcal{I}_2(\mathbf{x} + \mathcal{F}_0(\mathbf{x})) - \frac{\partial I_2(\mathbf{x} + \mathcal{F}_0(\mathbf{x}))}{\partial \mathbf{x}}\,(\mathcal{F}(\mathbf{x}) - \mathcal{F}_0(\mathbf{x})) \qquad (18)$$

The Euler Lagrange equation leads to the following diffusion equation:

$$\frac{\partial \mathcal{F}(\mathbf{x})}{\partial t} = \mathrm{div}(T(\nabla\mathcal{I}_1^*)\,\nabla\mathcal{F}(\mathbf{x})) - \lambda\mathcal{V}(\mathbf{x})\left(\frac{\partial \mathcal{I}_2(\mathbf{x} + \mathcal{F}_0(\mathbf{x}))}{\partial \mathbf{x}}\right)^T \mathbf{\Sigma}^{-1}\mathbf{m}_2(\mathbf{x}). \quad (19)$$

We now compare this result with two other important optical flow (OF) approaches, described in [2, 10]. First of all, the data term in eq.(19) now contains $\mathbf{\Sigma}^{-1}$, which performs a global, relative weighting of the different spectral components. Apart from this, $\mathbf{\Sigma}^{-1}$ also globally weights the importance of the matching term w.r.t. the smoothness term. More image noise decreases the norm of $\mathbf{\Sigma}^{-1}$. This automatically results in a

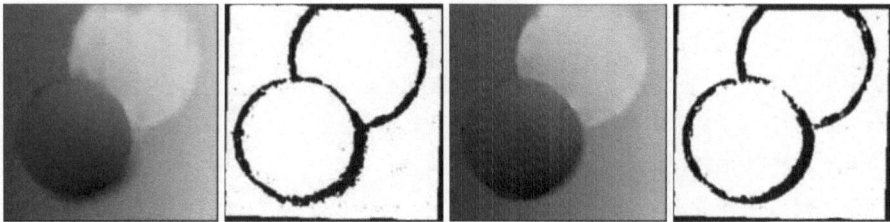

Fig. 3. Resulting optical flow fields and visibilities using the 2nd (left two images) and 3rd (right two images) algorithm, $log(\lambda) = -7$, histogram size $= 8^3$

more smooth solution, which is a desirable mechanism. A further modification of the data term is the local weighting of the image value differences by the visibilities $\mathcal{V}(\mathbf{x})$. When a pixel receives a low visibility score, the smoothness term locally gets more important. This avoids that wrongly matched occluded pixels pull, by the action of the smoothness term, neighboring unoccluded pixels in the wrong direction. Another difference is related to the model image $\mathcal{I}_1^*$. Instead of comparing $\mathcal{I}_1(\mathbf{x})$ with $\mathcal{I}_2(\mathbf{x}+\mathcal{F}_0(\mathbf{x}))$, which is the usual practice in OF computation, the Bayesian framework tells us to use $\mathcal{I}_1^*(\mathbf{x})$ instead. This results again in a visibility dependent weighting [2]. The algorithm has, similar to Alvarez *et al.*[2], two free parameters. They are ν, which controls the degree of anisotropy in eq. (11), and λ, which controls the width (hence the importance) of regularization prior. The smoothness term in eq. (19) is conceptually similar to the one used by Alvarez *et al.*. However, their regularizer is linear whereas ours is non-linear and $\mathcal{I}_1^*$ dependent.

3.2   Visibility Induced Anisotropy

In the second algorithm, we employ a smoothness term related to the work by Proesmans *et al.*[10], which outperformed other optical flow algorithms in a recent benchmark [8]. In [10], the basic idea is to compute optical flow from $\mathcal{I}_1$ to $\mathcal{I}_2$ and vice-versa, which are called forward and backward diffusion respectively. Ideally corresponding forward and backward flow vectors should sum up to zero. The consistency of the forward-backward match is measured by two other diffusion equations. These equations assume spatial smoothness of the consistency values and constraint them to the range $[0..1]$. Anisotropy is realized using these local consistency estimates. This approach needs optical flow computation in two directions (forward/backward) as well as the two other consistency diffusion equations with their own parameters. Instead of using these geometric consistencies, we propose to use the visibilities $\mathcal{V}(\mathbf{x})$ as a photometric consistency measure. Only the smoothness term in eq. (19) changes and we get:

$$\frac{\partial \mathcal{F}(\mathbf{x})}{\partial t} = \mathrm{div}(\mathcal{V}(\mathbf{x})\,\nabla\mathcal{F}(\mathbf{x})) - \lambda\mathcal{V}(\mathbf{x})\left(\frac{\partial \mathcal{I}_2(\mathbf{x}+\mathcal{F}_0(\mathbf{x}))}{\partial \mathbf{x}}\right)^T \mathbf{\Sigma}^{-1}\mathbf{m}_2(\mathbf{x}). \quad (20)$$

---

[2] Recall from eq. 16, that $\mathcal{I}_1^*(\mathbf{x}) - \mathcal{I}_2(\mathbf{x}+\mathcal{F}_0(\mathbf{x})) = (\mathcal{I}_1(\mathbf{x}) - \mathcal{I}_2(\mathbf{x}+\mathcal{F}_0(\mathbf{x})))/(1 + \mathcal{V}(\mathbf{x}))$.

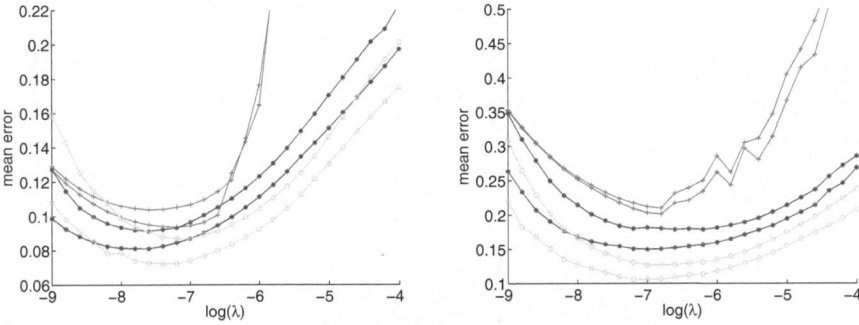

Fig. 4. Mean error of all visible pixels for three algorithms on two synthetic scenes (for 2 different amounts of Gaussian noise) as a function of λ. The different algorithms are shown in 1^{st}(red,+), 2^{nd}(blue,*), 3^{rd}(green,o)

This smoothness term blocks diffusion from places with high visibility estimates to places with lower ones. Typically, at initialisation time visibilities start out isotropically distributed while at the end pixels with low visibilities tend to cluster near discontinuities and occlusions. The algorithm has only one free parameter, being λ.

We will end with a final note on the convergence properties of the algorithms. Dempster *et al.* [6] have shown that, for Maximum-Likelihood (ML) estimation, each iteration of EM is guaranteed to increase the data-likelihood, which drives the parameters θ to a local optimum of L. In this work, we have included a prior on the unknown variables, so for the moment we can not make such strong claims. However, various trials on different data sets have confirmed the robust behavior of the two proposed algorithms.

4   Experiments and Discussion

We tested our two optical flow algorithms, together with our implementation of Alvarez *et al.*[2], on two synthetic scenes. To each of these scenes, different amounts of Gaussian noise were added. One of these scenes, together with the ground truth displacements and occlusions, is shown in fig. (2). The three evaluated algorithms are:

(1) image based anisotropic optical flow, our implementation of [2]
(2) image based anisotropic optical flow, with probabilistic matching term eq. (19)
(3) visibility based anisotropic optical flow, with probabilistic matching term eq. (20)

Fig. (3) shows the result of the 2nd and 3rd algorithm for the scene in fig. (2). The optical flow field is displayed color coded. We use the green channel for the horizontal and the blue channel for the vertical component of the flow field. The mean error (mean distance of ground truth with estimated optical flow) of all unoccluded pixels is shown in fig. (4) as a function of the λ parameter. The left hand plot of fig. (4) displays the results of the three algorithms on a scene with moving camera and a rotating and translating ball (not shown due to space limitations). We added Gaussian noise of variance 1 and 3 to the three color channels, so every algorithm appears twice in the plots. The

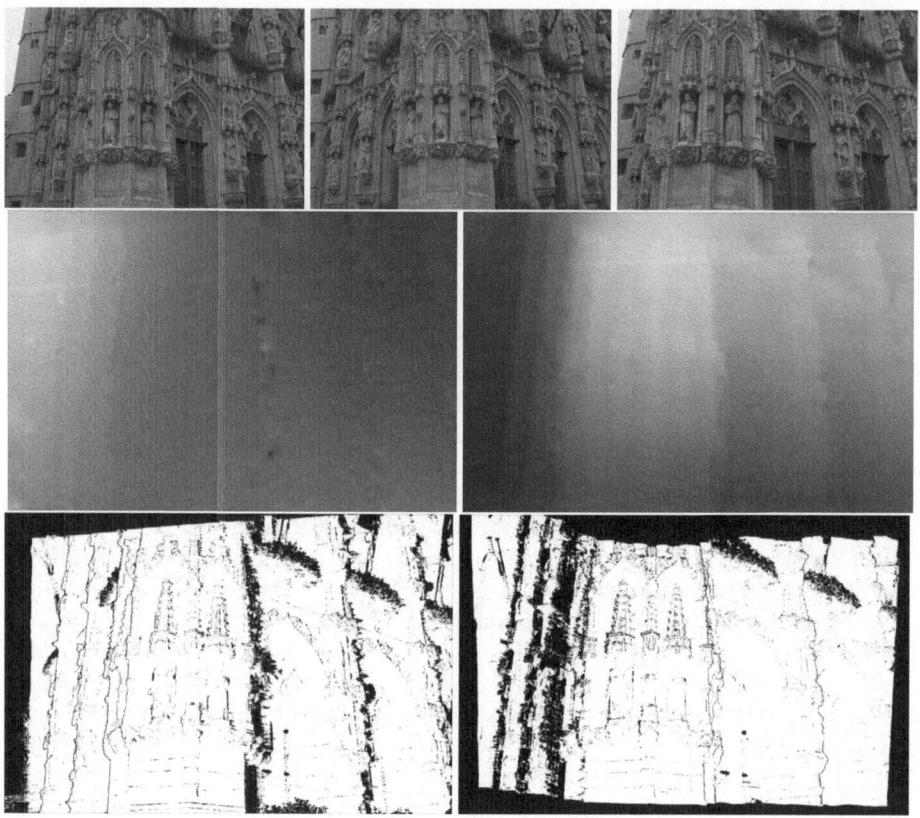

Fig. 5. City hall scene: top: input images $1, 2$ and 3; middle: independently computed optical flow from $\mathcal{I}_1$ to $\mathcal{I}_2$ (left) and from $\mathcal{I}_1$ to $\mathcal{I}_3$ (right), color coding as explained in the text; bottom: visibility maps for these two flow computations

right hand plot in fig. (4) displays the results on the scene shown in fig. (2) under the same experimental conditions. This scene is more complex than the first one.

In all synthetic experiments our probabilistic matching term performs better than the algorithm of Alvarez *et al.*[2]. Concerning the dependency on λ, one can see that Alvarez *et al.*(red,+) shows the expected behavior w.r.t. the amount of noise added. When the amount of noise increases, the optimal λ-value decreases, indicating the need for stronger regularisation. Our algorithms ((blue,*) and (green,o)) show an opposite behavior. This is probably due to the fact that the ML-estimate of Σ in eq. (16) under-estimates the true noise level when only two images are used. In conclusion, all methods have a noise-level dependent and scene dependent optimal λ. This dependency can only be resolved with prior knowledge of the complexity of the scene.

When comparing our two algorithms, one can observe better results for the 3rd algorithm, which incorporates the visibility based anisotropic smoothness term (eq. 20). The effect is even more outspoken when the complexity of the scene, i.e. the amount

Fig. 6. 3D reconstructions from optical flow estimation: left: result of the self-calibration; middle, right: two rendered un-textured views

of occlusions, increases. This result suggests that occlusions (visibilities) are estimated well in the probabilistic framework. This can also be observed from fig. (3) and fig. (5). By coupling the anisotropy to the visibilities instead of strong image gradients, especially nearby discontinuities, better flow estimates are obtained.

In two real data experiments, we show the results of our best performing 3rd algorithm. The city hall scene is characterized by large displacements and a strong impact of occlusions. Three images of resolution 3072×2048 have been used to compute optical flow and occlusions from image $\mathcal{I}_1$ to $\mathcal{I}_2$ and from image $\mathcal{I}_1$ to $\mathcal{I}_3$ (top images in fig. (5)) independently. The resulting optical flow fields and the visibility maps are show in fig. (5). Because we have no ground truth for these image pairs, we decided to evaluate the quality of the results by computing the 3D reconstruction. After processing the two image pairs we self-calibrated the scene using all optical flow correspondences from $\mathcal{I}_1$ to the other images. Only scene points with high visibility ($\mathcal{V}(\mathbf{x}) \geq 0.5$) to both views and with high spatial gradient were used. This results in the external and internal calibration of the cameras and a sparse set 3-D points. These are shown together with the camera positions on the left image in fig. (6). Using this camera calibration and the optical flow correspondences from $\mathcal{I}_1$ to $\mathcal{I}_2$ we computed the 3-D points, now for all pixels with high visibility. The result is shown as an untextured 3-D mesh rendered from two different virtual camera positions (right two images in fig. (6)). We wish to stress that this approach is purely for evaluation purpose. In real applications, one would of course restrict the correspondence search to the epipolar lines. However, even without epipolar constraints, the 2D optical flow matches result in a reliable 3D reconstruction, which is an indication for the quality of the computed flow field.

The visibility maps in fig. (5) show that the occluded pixels and discontinuities are accurately detected. Furthermore, they also show low visibility values for the specularities in the window closest to the viewer and for the flowers at the right top of the images. This is because specularities are classified as outliers given the local estimate of $\mathcal{I}_1^*$ and the global estimate of Σ. As a result, at these positions the importance of the matching term decreases and the flow field estimate is locally strongly regularized. Also pixel discretisation errors at high frequency detail (e.g. for the flowers) can bring about a similar effect.

5   Conclusions

We have presented two differential optical flow algorithms that simultaneously estimate image displacements and occlusions, as well as the noise distribution and denoised image, as part of an EM algorithm. Starting from relatively straightforward probabilistic considerations, we arrive at an energy formulation with a strong intuitive appeal. The energy often taken as a point of departure in other differential optical flow approaches, turns out to be a special case of this result. More specifically, it can be derived from our formulation by assuming unit strength noise, full visibility and by setting the unknown 'true' irradiance equal to the first image. The estimation of visibilities (occlusions) is naturally incorporated into the algorithm, similar in flavour to outlier detection in iteratively reweighted least squares estimation. This is a rather old concept, however, its use in optical flow computation and occlusion detection is new. Noticably, our best performing algorithm does not introduce additional parameters to realize anisotropic diffusion and to detect occluded pixels.

References

1. L. Alvarez, R. Deriche, T. Papadopoulo, and J. Snchez: Symmetrical dense optical flow estimation with occlusions detection. ECCV **1** (2002) 721–735
2. L. Alvarez, J. Weickert, and J. Snchez: Reliable estimation of dense optical flow fields with large displacements. IJCV **39** (2000) 41-56
3. G. Aubert, R. Deriche, and P. Kornprobst: Computing optical flow via variational techniques. SIAM Journal on Applied Mathematics **60(1)** (1999) 156–182
4. J. Barron, D. Fleet, and S. Beauchemin: Performance of optical flow techniques. IJCV **12(1)** (1994) 43–77
5. T. Brox, A. Bruhn, N. Papenberg, and J. Weickert: High accuracy optical flow estimation based on a theory for warping. ECCV **4**(2004) 25–36
6. A. P. Dempster, N. M. Laird, and D. B. Rubin: Maximum likelihood from incomplete data via the em algorithm. J. R. Statist. Soc. B **39** (1977) 1–38
7. B.K.P. Horn and B.G. Schunck: Determining optical flow. Artificial Intelligence **17**(1981)185–204
8. B. McCane, K. Novins, D. Crannitch, and B. Galvin: On benchmarking optical flow. Computer Vision and Image Understanding **84(1)** (2001) 126–143
9. H. H. Nagel: Constraints for the estimation of displacement vector fields from image sequences. Proc. Int. Joint Conf. Artificial Intell.(1983) 945–951
10. M. Proesmans, L. Van Gool, E. Pauwels, and A. Oosterlinck: Determination of optical flow and its discontinuities using non-linear diffusion. ECCV **2** (1994) 295–304
11. N. Vasconceles and A. Lippman: Empirical Bayesian EM-based Motion Segmentation. CVPR (1997) 527–532
12. J. Weickert and T. Brox: Diffusion and regularization of vector- and matrix-valued images. Inverse Problems, Image Analysis, and Medical Imaging. Contemporary Mathematics **313**(2002)

Mean-Shift Blob Tracking with Kernel-Color Distribution Estimate and Adaptive Model Update Criterion

Ning-Song Peng[1, 2] and Jie Yang[1]

[1] Institute of Image Processing and Pattern Recognition, Shanghai Jiaotong University,
P.O. Box 104, No.1954 Huashan Road, Shanghai, 200030, China
{pengningsong, jieyang}@sjtu.edu.cn
[2] Institute of Electronic and Information, Henan University of Science and Technology,
Luoyang, 471039, China

Abstract. We propose an adaptive model update mechanism for mean-shift blob tracking. It is novel for us to use self-tuning Kalman filters for estimating object kernel-color distribution, i.e. kernel-histogram. Filtering residuals are employed for hypothesis testing in order to obtain a robust criterion for model update. Therefore, tracker has the ability to keep up with the changes of object appearance as well as the changes in scale. Moreover, over-update is avoided in the cases of severe occlusion and dramatic appearance changes. Various tracking sequences demonstrate the superior behavior of our tracker which runs in real-time with non-parameter initialization and is robust to appearance changes.

1 Introduction

Visual tracking is a task required by various applications such as surveillance [1], video compression [2] and automatic video analysis [3]. An efficient and robust tracker should take real-time property into consideration as well as the ability to handle object changes, e.g. changes in object appearance or scale. Recently, the mean-shift algorithm is introduced to find target candidate that is most similar to a given model based on the measurement of Bhattacharyya coefficient [4,5]. It meets the real-time requirement without doing full search. However, the target model is not updated during the whole tracking period, which leads to poor localization when the object changes its scale or appearance. Although scale kernel is used in [6] for dealing with scaling problem, it has no ability to handle other appearance changes. To achieve robust performance, model update should be taken into more consideration. Over-update makes tracker sensitive to noises and occlusion, while under-update makes tracker lose the opportunities to keep up with the new object appearance. Updating model every frame based on the new matched position is not reasonable because it would induce over-update caused by accumulative matching errors. How to choose a tradeoff between the old model and the candidate model is a challenge work. Kalman

This work was supported by the National Natural Science Foundation of China under Grant No. 301702741.

D. Comaniciu et al. (Eds.): SMVP 2004, LNCS 3247, pp. 83–93, 2004.

filter meets our requirement because it is an optimal estimator by fusing information from measurement and previous states. In [7], each pixel in the template is allocated to a Kalman filter, which makes the template adaptive to the changes of object intensity. However, scaling problem limits its performance because this pixel-to-filter assumption holds no longer when object expands or shrinks its size. In this paper, we propose a new adaptive model update mechanism for real-time mean-shift tracking. Since the Kalman filter has been used in tracking mainly for smoothing the object trajectory [8,9,10], it is novel for us to use self-tuning Kalman filters for estimating object kernel-color distribution. In addition, hypothesis testing is employed as a criterion for determining whether to accept the estimate. Therefore, tracker has ability to avoid both under-update and over-update.

The paper is organized as follows. In Section 2, we first review the mean-shift theory, and then the model update mechanism is developed and analyzed. Experiment results are given in Section 3 and the conclusion is in Section 4.

2 Proposed Method

2.1 Mean-Shift Algorithm

The mean shift algorithm is an efficient and nonparametric method for nearest mode seeking [11,12] based on kernel density estimation (KDE). Let data be a finite set A embedded in the $n-$ dimensional Euclidean space X. The samples mean at $x \in X$ is

$$sm(x) = \frac{\sum_a K(a-x)w(a)a}{\sum_a K(a-x)w(a)}, \ a \in A \tag{1}$$

where K is a kernel function and w the weight. The difference $sm(x)-x$ is called mean-shift vector. The repeated movement of data points to the sample means is called the mean-shift algorithm. An important property of the mean-shift algorithm is that the local mean-shift vector computed at position x using kernel K points opposite to the gradient direction of the convolution surface

$$J(x) = \sum_a G(a-x)w(a) \tag{2}$$

K and G must satisfy the relationship

$$g'(r) = -ck(r), r = \|a-x\|^2, c > 0 \tag{3}$$

where g and k are profile of kernel G and K, respectively. A commonly used kernel is Gaussian kernel. Its profile is defined by

$$k(x) = \frac{1}{2\pi} \exp(-\frac{1}{2}\|x\|^2) \tag{4}$$

whose corresponding function g is also Gaussian-like.

In the mean-shift tracking algorithm [4,5], $J(x)$ is designed by using Bhat-tacharyya coefficient to measure the similarity of two kernel-histograms representing the object image and the candidate image, respectively.

2.2 Kernel-Histogram: Object Kernel-Color Distribution

In this paper, object model is represented by object kernel-color distribution, i.e. ker-nel-histogram. Let $\{x_i\}_{i=1,2,\cdots n}$ be the pixel locations of object image in a frame, cen-tered at position y. In the case of gray level image, the object kernel-histogram is defined by [4,5]

$$p_b = C\sum_{i=1}^{n} k(\left\|\frac{y-x_i}{h}\right\|^2)\delta[B(x_i)-b]\quad b=1\cdots m \tag{5}$$

p_b is the value of bin in the kernel-histogram with the index b and m is the number of bins. The set $\{p_b\}_{b=1,2,\cdots m}$ represents the object model. $B(x_i)$ is the quantization value of the pixel intensity at x_i. h is the kernel-bandwidth which normalizes coordi-nates of image so as to make radius of the kernel profile to be one. The constant C is derived by imposing the condition $\sum_{b=1}^{m} p_b = 1$.

2.3 Kernel-Histogram Filtering by Adaptive Kalman Filters

To achieve the stability and robustness of object tracking, lots of the tracking systems use the Kalman filter [8,9,10]. In most of their methods, the Kalman filter keeps track-ing of object changes in position and velocity, not the changes of object appearance. We use Kalman filter for filtering object kernel-histogram so as to obtain the optimal estimate of object model. It is well known that the performance of the Kalman filter depends on two key parts: the reasonable linear model and the initial parameters in-cluding the noise statistical properties and the initial state variance. It is impossible for us to have an accurate state equation and the noise statistical properties owing to the various changes in the scene. However, the object model we used is a kind of color distribution which is insensitive to noise and poor state equation. On the other hand, although it is hard to establish an accurate state equation, we can establish a more accurate observation equation with the help of mean-shift tracking algorithm which provides good localization with small matching error. The small matching error means the small measurement error, which makes it possible for us to estimate changeable variance of the state noise [7]. In conclusion, it is feasible to use Kalman filter for estimating object model.

At the convergence position provided by the mean-shift tracking algorithm, we can get a new kernel-histogram with the current kernel-bandwidth according to Eq. (5). We call it *observation model* while the kernel-histogram used in the mean-shift itera-tions for the current frame tracking is called *current model*. Kalman filters then yield a tradeoff between these two models and provide us an optimal estimate of the object model that is called *candidate model*, for the next frame tracking. In contrast to intensity filtering method [7], our method is insensitive to object scale change because the number of the filters, i.e. the number of bins in the kernel-histogram, doesn't change under any scale. Moreover, this feature leads to real-time computation even when the object expands its size.

Since each bin in the kernel-histogram changes independently, we can filter each of them independently by individual Kalman filter. Assume that i refers to the frame index and y_0^i the center point of the object blob in the frame i. Considering frame i, the state equation of each bin in the kernel-histogram is given by

$$p_b^i = p_b^{i-1} + w_{i-1} \qquad (6)$$

where p_b^i denotes the value of bin b, i.e. the state of bin b in the frame i and p_b^{i-1} is the counterpart in the frame $i-1$. w_{i-1} specifies the state noise which keeps changing owing to the factors such as the change of the object appearance, scale and the object orientation, etc. The estimation of the set $\{p_b^i\}_{b=1,2,\cdots m}$, i.e. $\{\hat{p}_b^i\}_{b=1,2,\cdots m}$ is the *candidate model* in the frame i, where m is the number of bins in the kernel-histogram. As common in Kalman filtering, w_{i-1} is assumed to be Gaussian distribution, and has the same variance σ_w^2 for all the bins in the kernel-histogram. The observation equation is given by

$$o_b^i = p_b^i + v_i \qquad (7)$$

where v_i is the measurement noise . Again we assume it has an identical Gaussian distribution for all the bins with the variance σ_v^2. $\{o_b^i\}_{b=1,2,\cdots m}$ is the *observation model* with respect to the convergence position y_0^i in the current frame i. y_0^i is obtained by using the mean-shift iterations as follows [4,5].

$$y_1^i = \frac{\sum_{j=1}^{n} x_j w_j g\left(\left\|\frac{y_0^i - x_j^i}{h}\right\|^2\right)}{\sum_{j=1}^{n} w_j g\left(\left\|\frac{y_0^i - x_j^i}{h}\right\|^2\right)} \qquad (8)$$

where $\{x_j^i\}_{j=1,2,\cdots n}$ is the set of pixel locations around its center y_0^i in the frame i and its size n is determined by the kernel-bandwidth h. w_j is the weight function defined by [4,5]

$$w_j = \sum_{u=1}^{m} \delta[B(x_j) - u] \sqrt{\frac{p_b^{i-1}}{p_b^i}} \qquad (9)$$

From the initial position y_0^i which is the same to the position y_0^{i-1} in the previous frame $i-1$, we use Eq. (8) to get a new position y_1^i in the current frame i. If $y_1^i \neq y_0^i$, let $y_0^i = y_1^i$ and continue to iterate until $y_0^i = y_1^i$. The final y_1^i is the new matched position, and let it be y_0^i again to prepare for the next frame tracking.

Now, considering two consecutive frames $i-1$ and i, for each bin b, we derive our model update equations according to Kalman filter theory.
State prediction equation:

$$\hat{p}_b^{i|i-1} = \hat{p}_b^{i-1} \tag{10}$$

Variance of prediction error:

$$q_b^{i|i-1} = q_b^{i-1} + \sigma_w^2 \tag{11}$$

where the superscript $i|i-1$ denotes the prediction for the frame i based on the information from previous frame $i-1$
Gain equation:

$$m_b^i = \frac{q_b^{i|i-1}}{q_b^{i|i-1} + \sigma_v^2} \tag{12}$$

Optimal estimation of p_b^i is

$$\hat{p}_b^i = \hat{p}_b^{i|i-1} + m_b^i(o_b^i - \hat{p}_b^{i|i-1}) \tag{13}$$

Variance of filtering error:

$$q_b^i = (1 - m_b^i)q_b^{i|i-1} \tag{14}$$

These equations require the following parameters to be known: the initial state variance q_b^0, the state noise variance σ_w^2 and the measurement noise variance σ_v^2. To tune these parameters, the filtering residual

$$r_b^i = o_b^i - \hat{p}_b^{i|i-1} \tag{15}$$

is used to set them [7], which suggests to comparing the estimated variance of the filtering residual with their theoretical variance [13]. At frame i, the variance of filtering residuals is estimated by averaging filtering residuals of all the bins over the last L frames:

$$(\bar{r}^i)^2 = \frac{1}{mL} \sum_{l=i-L+1}^{i} \sum_{b=1}^{m} (o_b^l - \hat{p}_b^{l|l-1})^2 \tag{16}$$

and then σ_w^2 is estimated by

$$\sigma_w^2 = (\bar{r}^i)^2 - \sigma_v^2 - \frac{1}{m}\sum_{b=1}^{m} q_b^{i-1} \tag{17}$$

Let $i = 1$, we can obtain initial parameters as follows [7].

$$\sigma_v^2 = 0.5(\bar{r}^1)^2, \quad \sigma_w^2 = 0, \quad q_\mu^0 = 0.5(\bar{r}^1)^2 \tag{18}$$

We make σ_v^2 unchanged in the whole tracking period owing to that the observation equation is credible not only in localization accuracy but also in noise tolerance. Tuning σ_w^2 is usually sufficient for the filters to adapt to the changes of object appearance. Now, all parameters are set automatically.

2.4 Hypothesis Testing: The Model Update Criterion

Although we now can update object model by kernel-histogram filtering, it is not suitable for us to accept the *candidate model* all the time. We should try to find a robust criterion to decide whether the *candidate mode* $\{\hat{p}_b^i\}_{b=1,2,\cdots m}$ should be accepted or not because over model update could make tracker sensitive to outliers. Filtering residuals are used to reject outliers [7]. In their method, when the residual exceeds a certain times its standard deviation according to an empirical threshold, the measurement is rejected and the state is not updated. Their threshold value comes from the experience not from the analysis of the dynamic scene. In this paper, we use filtering residuals as samples and then hypothesis testing is employed to make choice. Once we select a significance level α, the hypothesis testing depends less on residuals statistic properties and is robust to outliers.

Let us suppose the totality R is a normal distribution, $R \sim N(\mu, \sigma^2)$. μ and σ^2 are unknowns. Given the alternative hypothesis $H_1 : \mu > \mu_0$ and null hypothesis $H_0 : \mu \leq \mu_0$, we want to get the critical region of the hypothesis testing under the significance level α. Let $\{r_i\}_{i=1,2,\cdots N}$ be the samples from totality R. The number of sample is N. Since we don't know σ^2, we use

$$t = \frac{\bar{r} - \mu_0}{s / \sqrt{N}} \tag{19}$$

as the test statistic where s is the square root of the standard deviation which is a unbiased estimation of σ. If H_0 is true, then

$$t \sim t(N-1) \tag{20}$$

holds. Therefore, from the conditional probability

$$P(refuse\ H_0 | H_0\ is\ true) = P(\frac{\bar{r} - \mu_0}{s / \sqrt{N}} \geq cr) = \alpha \tag{21}$$

we can get $cr = t_\alpha(N-1)$, then the critical region is as follows

$$t = \frac{\bar{r} - \mu_0}{s/\sqrt{N}} \geq cr \tag{22}$$

From Eq. (16), μ_0 is set to $\bar{r}^i$, and the sample set is from $\{r_b^i\}_{b=1,2,\cdots m}$ which is the filtering residuals at current frame i. The number of the samples is m. Using Eq. (22), we can make choice between the two hypotheses. The acceptance of H_0 means it is necessary to accept *candidate model*, otherwise we remain using *current model*. Thus, Eq. (13) changes to

$$\hat{p}_b^i = \begin{cases} \hat{p}_b^{i|i-1} + m_b^i(o_b^i - \hat{p}_b^{i|i-1}) & \text{if accept } H_0 \\ \hat{p}_b^{i-1} & \text{if accept } H_1 \end{cases} \tag{23}$$

Where $\sum_{b=1}^m \hat{p}_b^i = 1$ should be held. Eq. (23) makes the tracker robust to occlusion and dramatic object appearance changes. Fig. 1. illustrates the outline of the whole tracking system.

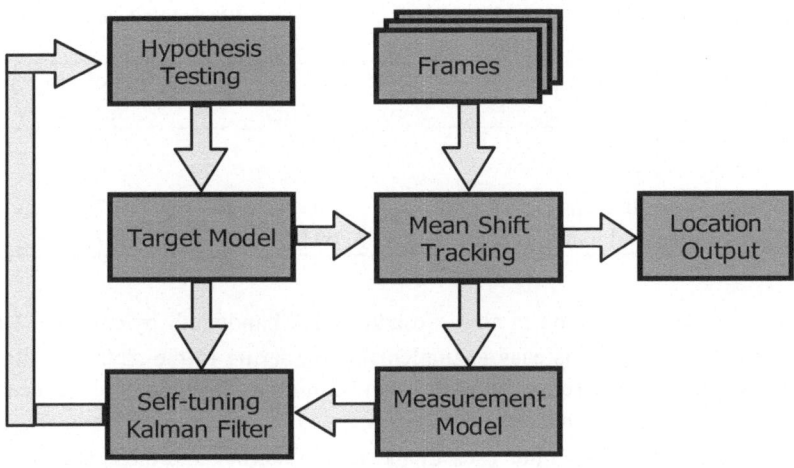

Fig. 1. The outline of the whole tracking system

3   Experimental Results

In our experiments, object kernel-histogram computed by Gaussian kernel has been derived in the *RGB* space with $32 \times 32 \times 32$ bins. Tab. 1 shows the statistics of the object kernel-histogram from *table tennis* sequence. We found that lots of bins are with the value zero. The reason lies in that the object color distribution is often so narrow in the natural scene. If filtering is done over all the bins, $\bar{r}^i$ would be near to zero. Consequently, σ_v^2 and $\{q_b^0\}_{b=1,2,\cdots m}$ maybe zero according to Eq. (17) and (18), which leads Kalman filters to be idle all the time. Since object color distribution doesn't change so much during the whole tracking period, we only process the bins with nonzero value, which doesn't impact the tracking accuracy and reduces the computation remarkably at the same time. Although some bins with zero value unprocessed by us may become nonzero when severe occlusion occurs, there must be also some corresponding large value changes of the bins that we processed all along, and the model update criterion could also work well. We start to setup initial parameters of Kalman filters after tracking 5-10 frames with $L = 3$. It is enough to get stable initial parameters by Eq. (17) and (18). Fig. 2. shows 4 frames with obvious object appearance changes in the *table tennis* sequences where the head of the player is our object and the white block shows the tracking result. Fig.3. shows the changes of σ_w^2 and the average of $\{q_b^i\}_{b=1,2,\cdots m}$. Every time, when the player drops down or raises his head, σ_w^2 rises up reflecting the object appearance changes, while, gradually, the average of $\{q_b^i\}_{b=1,2,\cdots m}$ comes to decrease reflecting that Kalman filters work well not only in estimating object model but also in tuning filter's parameters automatically.

Table 1. Statistics of the object kernel-histogram in frame 10 of *table tennis* sequence

object with RGB space of 32×32×32 bins	number of zero value bins	number of nonzero value bins
head	32644	124
left arm	32679	89
right arm	32697	71
bat and hand	32691	77

To track objects changing in size, we add kernel-bandwidth by $\pm \Delta h, \Delta h = 2$ to get the best blob scale [5]. It is easy to implement comparing to the method using scale space mean-shift [6], where the author uses an additional scale kernel to do mean-shift iterations in the scale space defined by $\{\delta_{scale} = \delta_0 1.1^{scale}, -2 \le scale \le 2\}$ in which δ_0 is the initial scale. Because the author uses Epanechnikov kernel, the mean-shift iterations for finding the best scale is equal to average all the scale in the scale space under the condition that the spatial space has been well matched. Obviously, this method is similar to our method except for its complexity. Fig.4. shows 4 frames of *traffic sequences*. The vehicle we tracked (white block) runs to the camera from far distance. The tracker performs well while object changes in size all the time and even warded off by another vehicle in frame 59.

We use significance level $\alpha = 0.10$ to do hypothesis testing. From the table of t distribution, we find that $t_{0.10}(n-1)$ changes little when $n > 10$. Therefore, it is enough to use 21 nonzero value bins from the kernel-histogram for hypothesis testing, i.e. $n = 21$. Fig. 5. shows 4 frames of *flower and garden* sequences with the attic as our object (white block). The dashed block shows the tracking result by directly using *candidate model* without the assistance of the hypothesis testing. When the full occlusion occurs, the tracker takes the tree as its target and loses the real target.

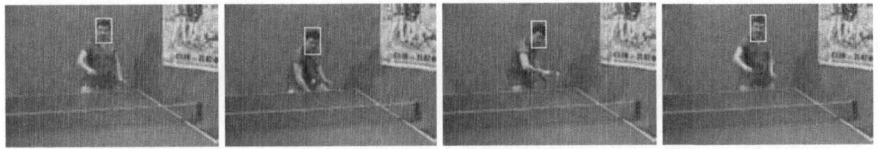

Fig. 2. *Table tennis* sequence: The frames 10, 26, 64, and 81 are shown (left-right)

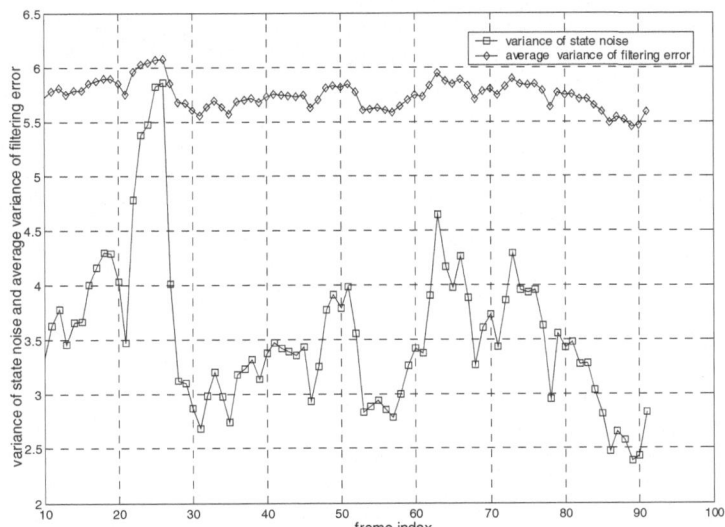

Fig. 3. Average of $\{q_b^i\}_{b=1,2,\cdots m} \times 10^6$ (diamond) and $\sigma_w^2 \times 10^5$ (square) in the *table tennis* sequence

Fig. 4. *Traffic* sequence: The frames 43, 59, 81, and 102 are shown (left-right)

The solid block shows the tracking result by using the full of our adaptive model update mechanism, and the target doesn't lose after the severe occlusion at frame 56. Fig. 6(a) shows the average filtering residuals of the last L frames, i.e. μ_0 (dot) and the average filtering residuals (cross) of the current frame. Fig. 6(b) shows $t_{0.10}(20)$ (flat line) and the test statistic curve. When full occlusion occurs at the 56th frame, test statistic t gets across the $t_{0.10}(20)$. This case doesn't take place in other frames, even though, in some frames, the average filtering residuals is bigger than μ_0, see also Fig. 6(a).

Fig. 5. *Flower and garden* sequence: The frames 10, 52, 56, and 66 are shown (left-right)

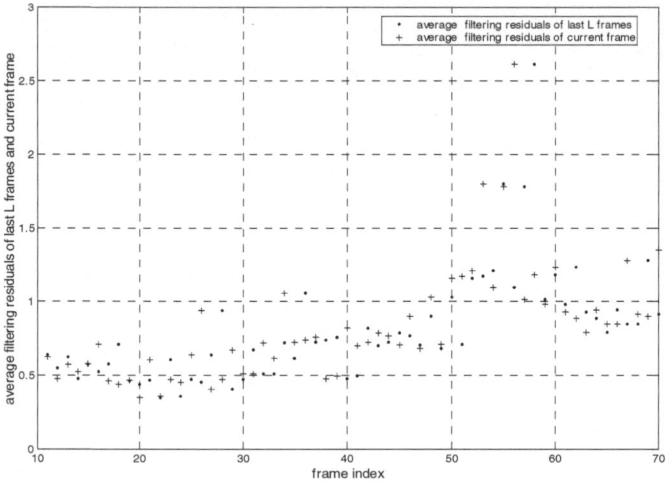

(a)

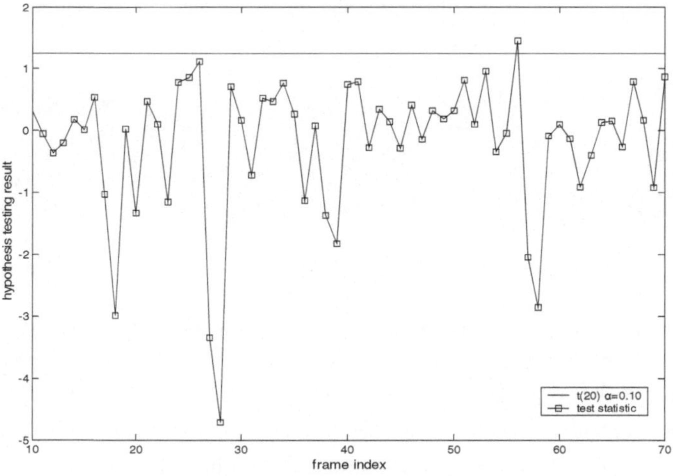

(b)

Fig. 6. (a) Average filtering residuals of current frame (cross) and μ_0 (dot) in the *flower and garden* sequence; (b) Hypothesis testing result of *flower and garden* sequence with $t_{0.10}(20) = 1.325$ (flat line)

4 Conclusion

This paper proposes a novel model update mechanism for the mean-shift based real-time object tracking. By using Kalman filter for filtering object kernel-histogram, the object model for the next frame tracking is continuously estimated. Therefore, the tracker has the ability to keep up with the changes of object appearance as well as the changes in size. Moreover, the tracker can handle severe occlusion and dramatic appearance changes in complexity scenes by the hypothesis testing to avoid over model update. With the mean-shift tracking algorithm and the self-tuning technique for constructing Kalman filters, the tracking process starts with no parameters initialization and the computational complexity satisfies the real-time requirement.

References

1. Cui Y., Samarasekera S., Huang Q., Greiffenhagen M.: Indoor monitoring via the collaboration between a peripheral sensor and a foveal sensor. IEEE Workshop on Visual Surveillance. (1998) 2-9
2. Eleftheriadis A., Jacquin A.: Automatic Face Location Detection and Tracking for Model-Assisted Coding of Video Teleconference Sequences at Low Bit Rates. Signal Processing-Image Communication. 3 (1995) 231-248
3. Wactlar H.D., Christel M.G., Gong Y., Hauptmann A.G.: Lessons learned from the creation and deployment of a terabyte digital video library. IEEE Computer. 2 (1999) 66-73
4. Comaniciu D., Ramesh V., Meer P.: Real-time tracking of non-rigid objects using mean shift. IEEE Int. Proc. Computer Vision and Pattern Recognition. 2 (2000) 142-149
5. Comaniciu D., Ramesh V., Meer P.: Kernel-based object tracking. IEEE Trans. Pattern Analysis Machine Intelligence. 5 (2003) 564-575
6. Collins R.T.: Mean shift blob tracking through scale space. IEEE Int. Proc. Computer Vision and Pattern Recognition. 2 (2003) 234-240
7. Nguyen H.T., Worring M., van den Boomagaard R.: Occlusion robust adaptive template tracking. IEEE Int. Conf. Computer Vision. 1 (2001) 678-683
8. Blake A., Curwen R., Zisserman A.: A framework for spatio-temporal control in the tracking of visual contour. Int. J. Computer Vision, 2 (1993) 127-145
9. Legters G., Young T.: A Mathematical Model for Computer Image Tracking. IEEE Trans. Pattern Analysis Machine Intelligence. 6 (1982) 583-594
10. Zhu Z., Ji Q., Fujimura K.: Combining Kalman filtering and mean shift for real time eye tracking under active IR illumination. IEEE Int. Conf. Pattern Recognition. 4 (2002) 318-321
11. Fukanaga K., Hostetler L.D.: The estimation of the gradient of a density function, with applications in pattern recognition. IEEE Trans. Information Theory. 1 (1975) 32-40
12. Cheng Y.: Mean shift, mode seeking and clustering. IEEE Trans. Pattern Analysis and Machine Intelligence. 8 (1995) 790-799
13. Maybeck P.: Stochastic models, estimation and control. Academic Press, New York(1982)

Combining Simple Models to Approximate Complex Dynamics

Leonid Taycher, John W. Fisher III, and Trevor Darrell

Vision Interface Group, CSAIL,
Massachusetts Institute of Technology, Cambridge, MA, 02139
{lodrion, fisher, trevor}@ai.mit.edu

Abstract. Stochastic tracking of structured models in monolithic state spaces often requires modeling complex distributions that are difficult to represent with either parametric or sample-based approaches. We show that if redundant representations are available, the individual state estimates may be improved by combining simpler dynamical systems, each of which captures some aspect of the complex behavior. For example, human body parts may be robustly tracked individually, but the resulting pose combinations may not satisfy articulation constraints. Conversely, the results produced by full-body trackers satisfy such constraints, but such trackers are usually fragile due to the presence of clutter. We combine constituent dynamical systems in a manner similar to a Product of HMMs model. Hidden variables are introduced to represent system appearance. While the resulting model contains loops, making the inference hard in general, we present an approximate non-loopy filtering algorithm based on sequential application of Belief Propagation to acyclic subgraphs of the model.

1   Introduction

One of the successful approaches to vision-based tracking is to formulate the problem as probabilistic inference in a dynamic model (e.g HMM or DBN). Since the exact underlying dynamics are either unknown or computationally intractable, standard approaches use approximate dynamics with inflated dynamical noise. Furthermore, the likelihood functions are often multi-modal, with some of the modes arising from the structures in image clutter (i.e., parts of observed images not generated by the object(s) of interest). The broad temporal priors combined with such multi-modal likelihoods can result in incorrect posterior estimates, since the broad prior is more likely to cover more than one peak in the likelihood.

It is well known that state representations, in general, are not unique. In this paper, we present a framework for combining individual dynamical models with *redundant* state representations and/or dynamics to improve state estimation. While the proposed framework is quite general, for the purpose of concreteness we will emphasize its application to tracking of articulated bodies (e.g. humans bodies).

A common approach to human-body tracking is to formulate the problem as filtering in a hidden Markov chain framework (cf. review in [10]), where an observation at a particular time step is the video frame, and the state is a particular representation of a

D. Comaniciu et al. (Eds.): SMVP 2004, LNCS 3247, pp. 94–104, 2004.

human body augmented with appropriate sufficient statistics of the past states. Most of these approaches use simple edge-correspondence-based image likelihood models, that are liable match to structured backgrounds such as bookshelves, etc.

Different choices are available for body representation. It can be described as a vector of joint angles of an appropriate articulated tree. Alternatively a set of 3D rigid poses of the constituent segments or a set of keypoints positions on the body may be used. Each representation has advantages and weaknesses: a set of joint angles (appropriately limited) always represents a valid body configuration, but complicated dynamic models are required for anything but relatively simple motions (such as walking and running); modeling body dynamics in the space of independent 3D rigid poses may be simpler [7], but such sets may describe invalid configurations.

Since individual models correctly model different aspects of the complex motion, they require dynamic noise models with different properties. The main intuition underlying our approach is that the differences in dynamic noise make them sensitive to different clutter structures. Visual structures corresponding to the physical system will be assigned high probability by *all* models, but different models may assign high probabilities to different structures in the clutter. Thus the state estimates can be made more robust to clutter by combining information from all available models and interpreting only structures to which all models assign significant prior probability, since these are likely to correspond to the system of interest.

Combining information from different models is not straightforward, since each model may use a different state representation. We address this issue by introducing a latent appearance representation, shared among all models, and combining models at this representation level. The interconnection allows information from one chain to influence the rest of the chains during the inference process, effectively serving as a data association filter. The resulting graphical model is loopy, which makes inference complicated in general, but we propose an approximate filtering algorithm for our framework that is based on sequentially applying Belief Propagation to acyclic subgraphs of the loopy model. We demonstrate that approximate filtering in the proposed multi-chain model applied to articulated-body tracking compares favorably with the standard CONDENSATION algorithm operating in a single-chain setting.

2   Prior Work

Object pose tracking formulated as inference in probabilistic frameworks has been a focus of research for the last decade. Most of the algorithms have been based on filtering in hidden Markov chain frameworks and share the "generate-and-test" method of observation likelihood computation. The likelihood is computed based on similarity between visual features corresponding to a particular state (produced by a deterministic function) and features extracted from image, and observation noise model. Various appearance features and 3D body descriptions have been proposed(cf. survey in [10]).

Multiple functional forms of the approximations to state posterior and state dynamics and likelihood distributions have been used. Early human tracking approaches [9] used a Kalman Filtering framework; thus implicitly modeling all constituent distributions with Gaussians. These assumptions have been later relaxed to account for nonlinear monomodal dynamics (while still using a Gaussian state model) by using Extended [5] or Unscented [13] Kalman Filters. Switched linear systems [8] were proposed to describe arbitrary learned dynamics.

Sample-based distribution representations have been used extensively in order to maintain multimodal state distributions. The original CONDENSATION algorithm and its variants were used in [11, 6]. The hybrid Monte Carlo sampling, which used observed image to modify sample locations, was presented in [2]. Partitioned [6] and Layered [14] sampling used factored state dynamics, allowing for semi-independent propagation of parts of the state vector. Several techniques (e.g. [12]) have been proposed for modifying state dynamics (or rather diffusion) parameters based on the previous state distribution. An implicit dynamics model presented by [10] uses a motion-capture database to predict the future state distribution.

Our approach is to combine simple cooperating trackers. In this sense it is similar to Bayesian modality fusion [16] – which combines the output of head trackers, operating in different modalities, based on individual trackers' reliability – and to Joint Bayes Filter [3] – which proposed decomposing hand motion into rigid and non-rigid motions – each of which is tracked with a separate Markov chain with reinforcing beliefs.

We propose combining simple trackers using the product model proposed in [4] and [1]. The major difference between our approach and Product of HMMs [1] is that the trackers are combined at the latent appearance rather than the observation level, which allows them to reinforce each other. Our model uses renormalized products of tractable probability distributions that are derived from individual trackers, each modeling a low-dimensional constraint on the data. In this work, we assume that constituent stochastic trackers are completely specified, and the appearance feature hierarchy (if any) is known. We concern ourselves with inference on the combined model, rather than learning its structure.

3   Multi-chain Model Formulation

Consider two models of human-body motion. One that parameterizes the pose $S_1^t \in \mathcal{S}_1$ as a vector of joint angles, and the other describing it as a set of rigid 3D poses of independently moving segments $S_2^t \in \mathcal{S}_2$ with approximate dynamics $\hat{p}(S_1^t|S_1^{t-1})$ and $\hat{p}(S_2^t|S_2^{t-1})$, and observation likelihoods $p(O^t|S_1^t)$ and $p(O^t|S_2^t)$, respectively (we use $\hat{p}(\cdot)$ to denote approximate functions).

Given the true dynamics, $p(S_i^t|S_i^{t-1})$, the state posterior distribution in each system ($i = 1, 2$) could be recursively computed with standard Bayesian recursion

$$p(S_i^t|O^{0..t}) \propto p(O^t|S_i^t)p(S_i^t|O^{0..t-1}), \qquad \text{where} \qquad (1)$$

$$p(S_i^t|O^{0..t-1}) = \int dS_i^{t-1} p(S_i^t|S_i^{t-1}) p(S_I^{t-1}|O^{0..t-1}). \qquad (2)$$

As has been stated, using approximate dynamics, $\hat{p}(S_i^t|S_i^{t-1})$ may cause multiple modes of the likelihood, $p(O^t|S_i^t)$, to be covered by the broad prior $\hat{p}(S_i^t|O^{0..t-1})$, and result in gross state misestimation. To mitigate this, we would like to incorporate the (approximate) prediction of other models, $\hat{p}(S_j^t|O^{0..t-1})$ into the posterior estimate

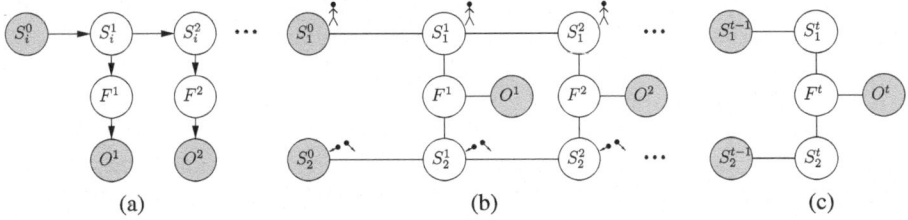

Fig. 1. Combining simple trackers. (a) An articulated body tracker with state $S_i^t \in \mathcal{S}_i$ (see text for details), dynamics $p(S_i^t|S_i^{t-1})$, feature-generation model $p(F^t|S_i^t)$ and feature-observation model $p(O^t|F^t)$. $p(O^t|S_i^t) = \int p(O^t|F^t)p(F^t|S^t)dF^t$. (b) Combined model with potentials corresponding to the conditional probabilities in the individual models. Different state spaces are indicated by graphics next to state nodes. (c) A tree-shaped subgraph on which a single step of approximate inference is performed. The marginal distributions, $p(S_1^{t-1}|O^{0..t-1})$ and $p(S_2^{t-1}|O^{0..t-1})$, have been computed at the previous iteration, and are not modified; O^t is observed

$\hat{p}(S_i^t|O^{0..t}), i \neq j$. To this end, we introduce the notion of a *common* latent appearance variable $F^t \in \mathcal{F}$ (Figure 1(a)). At each time step, this variable contains all information necessary to produce the observation from state S_i^t in *every individual model*. In our example, $\mathcal{F}$ may describe the 3D rigid-segment poses at the current time step, the edges in the images, etc. Since individual models describe the same observed image, such an $\mathcal{F}$ can always be selected.

Consider the joint distribution of the state and the appearance conditioned on all previous observations,

$$p(F^t, S_i^t|O^{0..t-1}) = p(F^t|S^t) \int dS_i^{t-1} p(S_i^t|S_i^{t-1}) p(S_i^{t-1}|O^{0..t-1}). \tag{3}$$

equality is not generally true for approximate dynamics,

$$p(F^t, S_i^t|O^{0..t-1}) = q(F^t, S^t; O^{0..t-1}) \int dS_i^{t-1} \hat{p}(S_i^t|S_i^{t-1}) p(S_I^{t-1}|O^{0..t-1}) \tag{4}$$

$$q(F^t, S^t; O^{0..t-1}) \neq p(F^t|S^t).$$

That is, when the true dynamics model is used, F^t (and O^t) are independent from prior observations conditioned on S_i^t, but this is not the case for the approximate dynamics. Modeling the dependency between F^t and prior observations that remains unmodeled by ith dynamic model would allows for better estimation of the ith state posterior. We choose the approximation to $q(F^t, S^t; O^{0..t-1})$ that incorporates the information from the other chain via a product

$$\hat{q}(F^t, S^t; O^{0..t-1}) \propto p(F^t|S_i^t) \int dS_j^t p(F^t|S_j^t) p(S_j^t|O^{0..t-1}), \quad i \neq j. \tag{5}$$

This captures the desired property that an individual model should consider only those values of appearance (i.e., structures) that are assigned significant prior probability by other models. The resulting system may be represented by the graphical model shown in Figure 1(b). The potentials in this undirected model are defined based on conditional distributions from constituent models (that is $\phi(S_i^t, S_i^{t-1}) = \hat{p}(S_i^t | S_i^{t-1})$, $\phi(F^t, S_i^t) = p(F^t | S_i^t)$, $\phi(O^t, F^t) = p(O^t | F^t)$).

Although we discuss the case when individual models use the same latent appearance features, it is possible to combine models with intersecting feature sets. In that case, the combined feature model would be the union of individual feature sets, and the likelihood potentials are extended to produce uniform likelihoods for features that are not part of an original submodel.

3.1   Inference in the Multi-chain Model

Single-chain models are popular because there exist efficient inference algorithms for them. While our proposed multi-chain model is loopy (Figure 1(b)) and exact inference is complicated in such models, we take advantage of the fact that we are interested only in marginal distributions for the state nodes to propose an efficient algorithm for *filtering* in our multi-chain model.

Consider the model in Figure 1(b). At time $t = 1$, we are concerned with nodes with superscripts (times) $t \leq 1$. If the initial states S_1^0 and S_2^0 are independent (as shown), then the resulting subgraph is a tree, and we can use standard Belief Propagation technique to compute exact marginal distributions at state nodes S_1^1 and S_2^1.

$$p(S_1^1 | O^1) = \frac{1}{Z} \int dS_1^0 \phi(S_1^1, S_1^0) p(S_1^0) \int dF^1 \left[\phi(F^1) \phi(F^1, S_1^1) \right. \tag{6}$$
$$\left. \int_{S_2^1} \left[\phi(F^1, S_2^1) \int S_2^0 \phi(S_2^1, S_2^0) p(S_2^0) \right] \right],$$

where $\phi(F^1) = \phi(O^1, F^1)$ (the equivalent expression of $p(S_2^1 | O_1)$ is not shown).

Filtering at the next timestep ($t = 2$) is more complex since the model now contains loops and the exact inference would require representing the joint $p(S_1^1, S_2^1 | O^1)$:

$$p(S_1^2 | O^1, O^2) = \frac{1}{Z} \int dF^2 \left[\phi(F^2) \phi(F^2, S_1^2) \int dS_2^2 \left[\phi(F^2, S_2^2) \right. \right. \tag{7}$$
$$\left. \left. \int dS_1^1 dS_2^1 \phi(S_1^2, S_1^1) \phi(S_2^2, S_2^1) p(S_1^1, S_2^1 | O^1) \right] \right]$$

In order to simplify computations, we approximate the joint distribution, $p(S_1^1, S_2^1 | O^1)$ with a product, $q(S_1^1) q(S_2^1)$. It is easily shown that the best such approximation (in the KL-divergence sense) is the product of marginal distributions, $p(S_1^1 | O^1)$ and $p(S_2^1 | O^1)$. Substituting $p(S_1^1 | O^1) p(S_2^1 | O^1)$ for $p(S_1^1, S_2^1 | O^1)$ in Equation 7, we obtain an approximate inference equation:

$$p(S_1^2 | O^2) \approx \frac{1}{Z} \int dS_1^1 \phi(S_1^2, S_1^1) p(S_1^1 | O^1) \int dF^2 \left[\phi(F^2) \phi(F^2, S_1^2) \right. \tag{8}$$
$$\left. \int dS_2^2 \left[\phi(F^2, S_2^2) \int dS_2^1 \phi(S_2^2, S_2^1) p(S_2^1 | O^1) \right] \right].$$

The similarity between Equations (6) and (8) suggests an approximate filtering algorithm that estimates marginal distributions of the state variables by recursively applying Belief Propagation to acyclic subgraphs of the form shown in Figure 1(c), using the marginal state distribution obtained at time $t - 1$ as priors at time t.

It may be shown that this approximation preserves the main property of the exact model: the appearance features that are assigned zero probability under *any* of the constituent models are assigned zero probability in computation of *all* of the marginal distributions.

At each timestep, an approximate inference may be performed using message passing in 4 steps.

1. Compute $\mu_{S_i^{t-1} \to S_i^t} = \int dS_i^{t-1} \phi(S_i^t, S_i^{t-1}) p(S_i^{t-1}|O^{0..t-1}), i = 1, 2$
2. Compute $\mu_{S_i^t \to F^t} = \int dS_i^t \phi(F^t, S_i^t) \mu_{S_i^{t-1} \to S_i^t}, i = 1, 2$
3. Compute $\mu_{F^t \to S_i^t} = \int dF^t \mu_{S_{j \neq i}^t \to F^t} \phi(O^t, F^t), i = 1, 2$
4. Compute marginal state distributions $p(S_i^t|O^{0..t}) \propto \mu_{S_i^{t-1} \to S_i^t} \mu_{F^t \to S_i^t}. i = 1, 2$

If inference on constituent Markov chains were performed individually, it would still involve steps analogous to 1 and 4 (and partially 3); consequently, combining models introduces very little additional complexity to the inference process.

4   Dual-Chain Articulated Tracking

We have used the multi-chain framework proposed in this paper for tracking human upper-body motion. As discussed in the section 3, we track the body using two different body descriptions. One is a commonly used articulated tree model with 13 degrees of freedom: 2 in-plane translational dofs (for the purposes of this application we assume that the body is observed under orthogonal projection, and the person is facing the camera), 3 rotational dofs at the neck, 3 rotational dofs at each shoulder and 1 rotational dof at each elbow.

Rather than common particle-filtering approaches (e.g., CONDENSATION), which use the sampled representation of the posterior at the previous frame to guide current-frame sample placement, we use the *likelihood sampling* approach proposed in [15]. The pose samples are drawn from the pose likelihood function and are then reweighted based on the propagated prior. Although this method results in greater per-sample complexity, it enables us to use many fewer samples, since they are in general placed more appropriately with respect to the posterior distribution (the likelihood function is sharper than the approximate temporal prior, and thus is a better proposal distribution).

To implement likelihood sampling, we take an advantage of the fact that we are able not only to evaluate, but also to sample from observation likelihood for the head and hands (in this case mixtures of Gaussians corresponding to face detector outputs and to detected flesh-colored blobs). We define observation likelihood using latent image observation likelihoods: face detector output for the head segment, flesh-color likelihoods for the hands, and occlusion edge map matching for the rest of the segments. Once the 2D face and hand position samples has been drawn, we use them together with inverse kinematics constraints to define pose-proposal distribution. This distribution is then used in the importance sampling framework to obtain sample from the pose likelihood. The complete description of likelihood pose-sampling may be found in [15].

The second chain used in our framework is a low-level hand and face tracker using flesh-blob detection and face detection as observations in a robust Kalman filter setup. The second chain is necessary since, while a tracker based on the likelihood sampling can successfully operate with a small number of samples and is self recovering, it is extremely sensitive to feature-detector failures (such as flesh-color misdetections). The appearance features shared between chains are hand and face positions, and the chains are combined in the manner described in Section 3.

We have applied our dual-chain tracker to three sample sequences, with results shown in Figure 2. For each frame in the sequence we have rendered fourty randomly drawn samples from the posterior state distribution (the frontal view overlayed on top of the input image is shown in the middle row, and side view is shown in the bottom row). The tracking results for the first sequence are also available in the submitted video file (rendered at one third of the framerate). In most frames, the tracker succeeded in estimating poses (that contained significant out of plane components and self occlusions), and was able to recover from mistracks (e.g. around frame 61 in the third sequence).

In Figure 3, we compare the performance of the dual-chain tracker described above using 1000 samples per frame (first column), the likelihood-sampling tracker using 1000 samples (second column), the CONDENSATION tracker with 5000 samples (which runs as fast as the dual-chain tracker) (third column), and finally the CONDENSATION tracker with 15,000 samples (the smallest number of samples that enables CONDENSATION to perform with accuracy approaching dual-chain tracker performance).The results are presented using the same method as in Figure 2, the frontal view is shown overlayed on top of the input image, side view is shown to the right. For the purposes of evaluation, we consider the tracker to have failed on a particular frame if none of the drawn particles correspond to the pose.

The dual-chain tracker was able to track the body with a 3% failure rate. The likelihood-sampling tracker was generally able to correctly estimate the pose distribution, but failed on 25% of the frames where image features were not correctly extracted (e.g., frame 20). The CONDENSATION variant with 5000 samples failed after just 30 frames due partly to sample impoverishment. Increasing the size of sample set to 15,000 (with similar increase of the running time) allowed CONDENSATION to track through most of the sequence with a 20% failure rate before completely loosing track.

5   Conclusions and Discussion

We have proposed a methodology for combining simple dynamical models with redundant representations as a way of modeling more complex dynamical structures such as a moving human body. The approach was motivated by the simple observation that nearly all "generate-and-test" approaches to tracking complex structures implicitly marginalize over an intermediate feature representation between state and observation. By making the feature representation explicit in our approach we obtained a straightforward means of mediating between simpler models as a means of capturing more complex behavior.

Sequence 1

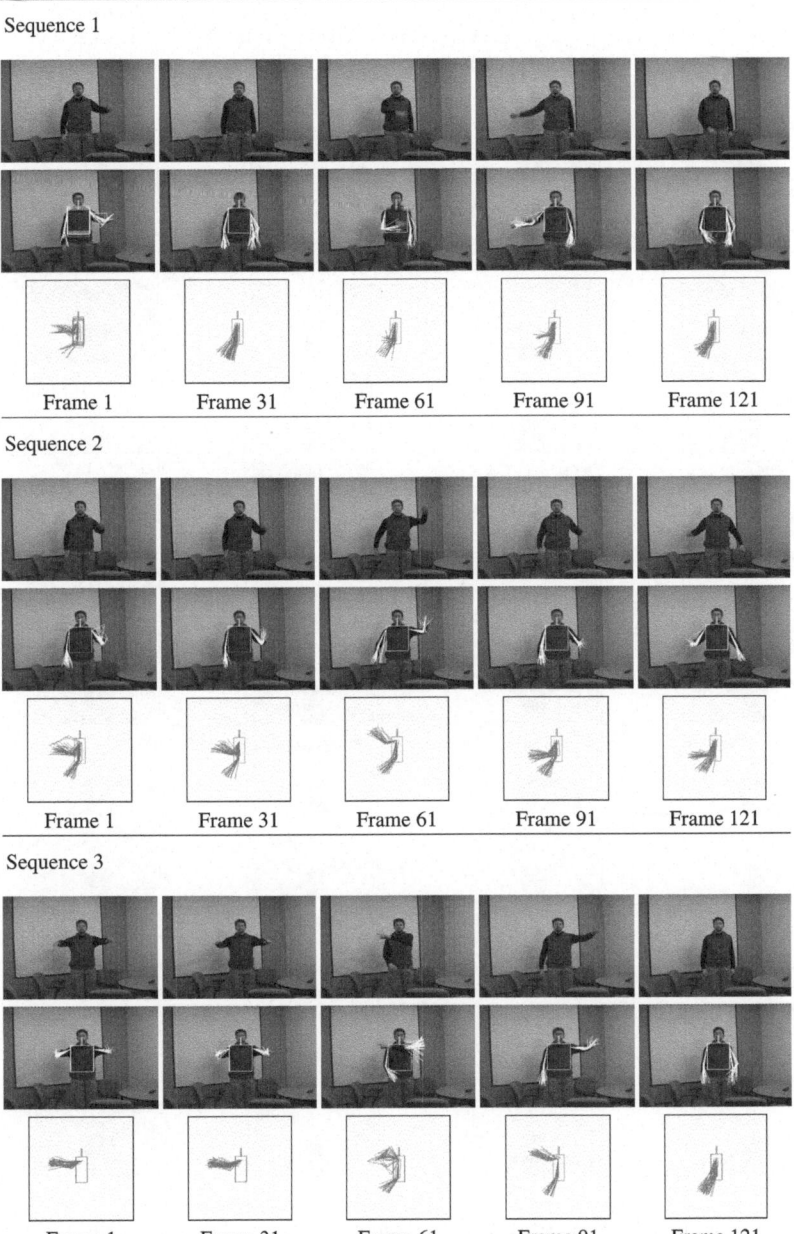

Sequence 2

Sequence 3

Fig. 2. Applying dual-chain tracking to three sample sequences. Five frames from each sequence are presented. The top row contains input frames. Fourty random particles from the estimated posterior pose distributions are shown: in the middle row, the patricles are rendered onto the input image (frontal view), and in the bottom row they are rendered in the side view. Note that while a mistrack has occured on the third sequence near frame 61, the tracker was able to recover

Fig. 3. Applying four tracking algorithms to a sample sequence. For each frame a set of fourty random pose samples were drawn from estimater posterior distribution and the corresponding skeletons was rendered (frontal view overlayed on the frame and side view next to it). Errors in feature detection caused likelihood-sampling tracker to fail on some of the frames (no samples were produced)

Exact inference on the resulting structure is complicated due to the introduction of loops in the graphical structure representing the combined models. However, as a consequence of the fact that we are primarily interested in the filtering (or tracking) problem, rather than the smoothing problem, an approximate inference method, based on sequential inference on acyclic subgraphs provides a suitable alternative to exact inference. This approximation has the important property that infeasible configurations in *any* of the naive models precluded an infeasible configuration in *all* of the others.

Empirical results demonstrated the utility of the method for tracking the upper body of a human. The method compares favorably with the well-known CONDENSATION algorithm in two ways. First, a monolithic approach using CONDENSATION required a significantly greater number of samples in order to explore the configuration space sufficiently as compared to the multi-chain method. Secondly, and perhaps more importantly, in the experiments presented the estimate of the posterior state distribution more accurately represents the uncertainty of the upper-body pose than the alternative methods. This is particularly encouraging considering the simplicity of combining constituent models as compared to a monolithic approach.

References

1. Andrew Brown and Geoffrey E. Hinton. Products of hidden markov models. In *Proceedings of Artificial Intelligence and Statistics*, pages 3–11, 2001.
2. Kiam Choo and David J. Fleet. People tracking using hybrid monte carlo filtering. In *Proc. ICCV*, 2001.
3. Huang Fei and Ian Reid. Joint bayes filter: A hybrid tracker for non-rigid hand motion recognition. In *ECCV (3)*, 2004.
4. Geoffrey E. Hinton. Products of experts. In *Proc, of the Ninth International Conference on Artificial Neural Networks*, pages 1 – 6, 1999.
5. Nebojsa Jojic, Matthew Turk, and Thoman S. Huang. Tracking self-occluding articulated objects in dense disparity maps. In *Proc of International Conference on Computer Vision*, 1999.
6. John MacCormick and Michael Isard. Partitioned sampling, articulated objects, and interface-quality hand tracking. In *ECCV (2)*, pages 3–19, 2000.
7. P. Morasso. Spatial control of arm movements. *Experimental Brain Research*, 42:223–227, 1981.
8. Vladimir Pavlović, James M. Rehg, Tat-Jen Cham, and Kevin P. Murphy. A dynamic bayesian network approach to figure tracking using learned dynamic models. In *Proc. ICCV*, 1999.
9. K. Rohr. Towards models-based recognition of human movements in image sequences. *CVGIP*, 59(1):94–115, Jan 1994.
10. Hedvig Sidenbladh. *Probabilistic Tracking and Reconstruction of 3D Human Motion in Monocular Video Sequences*. PhD thesis, Royal Institute of Technology, Stockholm, 2001.
11. Hedvig Sidenbladh, Michael J. Black, and David J. Fleet. Stochastic tracking of 3d human figures using 2d image motion. In *Proc. European Conference on Computer Vision*, 2000.
12. Christian Sminchiesescu and Bill Triggs. Covariance scaled sampling for monocular 3d human tracking. In *Proc. IEEE Conf. on Computer Vision and Pattern Recognition*, 2001.
13. B. Stenger, P. R. S. Mendonca, and R. Cipolla. Model-based hand tracking using an unscented kalman filter. *Proc. British Machine Vision Conference*, 2001.

14. J. Sullivan, Andrew Blake, Michael Isard, and John MacCormick. Object localization by bayesian correlation. In *ICCV (2)*, pages 1068–1075, 1999.
15. Leonid Taycher and Trevor Darrell. Bayesian articulated tracking using single frame pose sampling. In *Proc. 3rd Int'l Workshop on Statistical and Computational Theories of Vision*, Oct 2003.
16. Kentaro Toyama and Eric Horvitz. Bayesian modality fusion: Probabilistic integration of multiple vision algorithms for head tracking. In *ACCV'00*, 2000.

Online Adaptive Gaussian Mixture Learning
for Video Applications

Dar-Shyang Lee

Ricoh California Research Center,
2882 Sand Hill Road, Menlo Park, CA 94025, USA
dsl@rii.ricoh.com

Abstract. This paper presents an online EM learning algorithm for training adaptive Gaussian mixtures for non-stationary video data. Existing solutions are either slow in learning or computationally and storage inefficient. Our solution is derived based on sufficient statistics of the short-term distribution. To avoid unnecessary computation or storage, we show that the equivalent estimates can be accomplished by a set of recursive parameter update equations with one additional variable. The solution is evaluated against several existing algorithms on both synthetic data and surveillance videos. The results showed remarkable learning efficiency and robustness over current solutions.

1   Introduction

Adaptive Gaussian mixtures are becoming popular for modeling the temporal distribution of video data because of their compact and analytical representation. The nature of many vision applications requires the models to be adaptive over time and learning to be performed online. This prevents direct application of the traditional batch learning by Expectation Maximization [1] or incremental EM [7] developed for stationary distributions. The most common solution today for online adaptive mixture learning for non-stationary distributions employs exponential decay through recursive filtering to achieve temporal adaptability [8]. However, this is notoriously slow in convergence. Other solutions either require additional computation and storage overhead for buffering statistics in the adaptive window [6] or provide approximations without good theoretical foundation [4],[5].

In this paper, we present an online EM learning algorithm for non-stationary distributions formulated on short-term sufficient statistics inside the temporal adaptive window. Under explicit assumptions, we show that a set of equivalent recursive parameter update equations can be derived that has practically no storage or computational overhead. We compared the derived optimal learning algorithm with previously proposed methods on both synthetic and real video data. Experimental results showed excellent robustness and efficiency compared to other methods.

The rest of the paper is organized as follows. In Section 2, we formulate the online adaptive mixture learning problem and present the derivation of the learning equations. A comparative analysis of our derived solution against other related work is provided in

D. Comaniciu et al. (Eds.): SMVP 2004, LNCS 3247, pp. 105–116, 2004.

Section 3. We evaluate several algorithms using large simulation as well as real video data. Experimental results are presented in Section 4, followed by conclusions.

2   Online EM Learning for Adaptive Gaussian Mixtures

In vision applications, adaptive Gaussian mixtures offer a compact and analytical representation of input data distribution. In order to work over a long duration, it must be able to adapt to condition changes.

The problem can be stated as follows. We would like to model the short-term distribution of the intensity values observed at a single pixel location in the video over time using a Gaussian mixture. The model must be adaptive to reflect the non-stationary nature of the process, and learning must be done online without a presumptuous storage requirement.

2.1   Case 1: O(*LK*) Solution

Let $\mathbf{x}_1..\mathbf{x}_n$ be a sequence of data points whose temporal distribution we are trying to approximate by a mixture density function of K Gaussian components

$$P(\mathbf{x}) = \sum_{j=1}^{K} G(\mathbf{x};\theta_j) = \sum_{j=1}^{K} w(j) \cdot g(\mathbf{x};\mu(j),\Sigma(j)) \text{ and } \sum_{j=1}^{K} w(j) = 1 \tag{1}$$

where $g(x;\mu,\Sigma)$ is a single normal distribution. Since the assignment of the Gaussian components is not directly observed, maximum likelihood parameter estimates are obtained through an iterative Expectation-Maximization procedure [1]. For stationary distributions, stochastic versions such as the incremental EM algorithm [4] have been shown to achieve the same result by computing a set of sufficient statistics.

Let $q_n(j)$ be the expected posterior probability of component G_j for data point $\mathbf{x}_n$

$$q_n(j) \equiv P(G_j \mid \mathbf{x}_n) = \frac{w_n(j) \cdot g(\mathbf{x}_n;\mu_{n-1}(j),\Sigma_{n-1}(j))}{P(\mathbf{x}_n)}, \tag{2}$$

and let $Q_n(j)$, $M_n(j)$, $V_n(j)$ be sufficient statistics defined as

$$Q_n(j) \equiv \sum_{i=1}^{n} q_i(j)$$
$$M_n(j) \equiv \sum_{i=1}^{n} q_i(j) \cdot \mathbf{x}_i \tag{3}$$
$$V_n(j) \equiv \sum_{i=1}^{n} q_i(j) \cdot (\mathbf{x}_i - \mu_i(j))(\mathbf{x}_i - \mu_i(j))^T.$$

Then mixture parameters are computed by

$$w_n(j) = Q_n(j) / \sum_{k=1}^{K} Q_n(k)$$
$$\mu_n(j) = M_n(j) / Q_n(j) \tag{4}$$
$$\Sigma_n(j) = V_n(j) / Q_n(j).$$

In order for the model to adapt to distribution changes, we must discount older statistics as more samples are observed. To model the short-term distribution within a window of L recent samples $\mathbf{x}_{n-L+1}..\mathbf{x}_n$, the sufficient statistics for data in that window can be computed by

$$
\begin{aligned}
Q_n^L(j) &\equiv Q_n(j) - Q_{n-L}(j) \\
M_n^L(j) &\equiv M_n(j) - M_{n-L}(j) \\
V_n^L(j) &\equiv V_n(j) - V_{n-L}(j).
\end{aligned}
\tag{5}
$$

Based on these quantities, similar equations as Eq.(4) can be formulated to compute the parameters for an adaptive mixture model

$$
\begin{aligned}
w_n^L(j) &= Q_n^L(j) / \sum_{k=1}^{K} Q_n^L(k) \\
\mu_n^L(j) &= M_n^L(j) / Q_n^L(j) \\
\Sigma_n^L(j) &= V_n^L(j) / Q_n^L(j).
\end{aligned}
\tag{6}
$$

However, computing Eq.(5) would require storing sufficient statistics for all points within the most recent L-window, or O(LK) amount of storage, which is prohibitive in video applications. We would like to develop an approximation that is more efficient.

2.2 Case 2: O(K) Approximation

Since we are computing the statistics within a sliding window, we can rewrite Eq.(5) recursively to get

$$
Q_n^L(j) \equiv Q_{n-1}^L(j) - q_{n-L}(j) + q_n(j)
\tag{7}
$$

$$
M_n^L(j) \equiv M_{n-1}^L(j) - q_{n-L}(j) \cdot \mathbf{x}_{n-L} + q_n(j) \cdot \mathbf{x}_n
\tag{8}
$$

$$
\begin{aligned}
V_n^L(j) &\equiv V_{n-1}^L(j) - q_{n-L}(j) \cdot (\mathbf{x}_{n-L} - \mu_{n-L}^L(j))(\mathbf{x}_{n-L} - \mu_{n-L}^L(j))^T \\
&\quad + q_n(j) \cdot (\mathbf{x}_n - \mu_n^L(j))(\mathbf{x}_n - \mu_n^L(j))^T.
\end{aligned}
\tag{9}
$$

Considering that the distribution is quasi-stationary, it is reasonable to assume $q_{n-L}(j) \approx Q_{n-1}^L(j)/L$, then Eq.(7) simplifies to

$$
Q_n^L(j) \equiv Q_{n-1}^L(j) - q_{n-L}(j) + q_n(j) \approx \tfrac{L-1}{L} Q_{n-1}^L(j) + q_n(j).
\tag{10}
$$

We can make similar assumptions to simplify M_n^L and V_n^L. Assuming $q_{n-L}(j)\mathbf{x}_{n-L} \approx M_{n-1}^L(j)/L$, then Eq.(8) simplifies to

$$
M_n^L(j) \approx \tfrac{L-1}{L} M_{n-1}^L(j) + q_n(j) \cdot \mathbf{x}_n.
\tag{11}
$$

We further assume $q_{n-L}(j) \cdot (\mathbf{x}_{n-L} - \mu_{n-L}^L)^T (\mathbf{x}_{n-L} - \mu_{n-L}^L) \approx V_{n-1}^L(j)/L$, so

$$V_n^L(j) \approx \tfrac{L-1}{L} V_{n-1}^L(j) + q_n(j) \cdot (\mathbf{x}_n - \mu_n^L(j))(\mathbf{x}_n - \mu_n^L(j))^T . \tag{12}$$

From Eq.(10),(11),(12) the adaptive mixture parameters can be computed using Eq.(6). The amount of extra storage required in addition to mixture parameters is reduced to O(K). However, for video applications where every pixel is represented by a mixture model, maintaining and updating both sufficient statistics and model parameters, especially the vector manipulations involving M_n^L and V_n^L, still represent a significant overhead in storage and computation. We can eliminate the need for storing M_n^L and V_n^L from the following derivation.

Using Eq.(10), the weight calculation in Eq.(6) can be written as

$$w_n^L(j) = [\tfrac{L-1}{L} Q_{n-1}^L(j) + q_n(j)] / L = \alpha \cdot w_{n-1}^L(j) + (1-\alpha) \cdot q_n(j) \tag{13}$$

where $\alpha = (L-1)/L$ controls the rate of adaptation. Substituting Eq.(10) and (11) into Eq.(6), the recursive equation for computing the mean can be derived

$$\mu_n^L(j) = \frac{\tfrac{L-1}{L} M_{n-1}^L(j) + q_n(j) \cdot \mathbf{x}_n}{\tfrac{L-1}{L} Q_{n-1}^L(j) + q_n(j)} = \frac{\alpha \cdot \mu_{n-1}^L(j) + \frac{q_n(j)}{Q_{n-1}^L(j)} \cdot \mathbf{x}_n}{\alpha + \frac{q_n(j)}{Q_{n-1}^L(j)}} \tag{14}$$

$$= (1 - \eta_n^L(j)) \cdot \mu_{n-1}^L(j) + \eta_n^L(j) \cdot \mathbf{x}_n$$

where the learning rate η can be simplified using Eq.(10)

$$\eta_n^L(j) = \frac{\frac{q_n(j)}{Q_{n-1}^L(j)}}{\alpha + \frac{q_n(j)}{Q_{n-1}^L(j)}} = \frac{q_n(j)}{\alpha \cdot Q_{n-1}^L(j) + q_n(j)} = \frac{q_n(j)}{Q_n^L(j)} . \tag{15}$$

Similarly, substituting Eq.(10) and (12) into the variance update equations results in

$$\Sigma_n^L(j) = \frac{\tfrac{L-1}{L} V_{n-1}^L(j) + q_n(j)(\mathbf{x}_n - \mu_n^L(j))(\mathbf{x}_n - \mu_n^L(j))^T}{\tfrac{L-1}{L} Q_{n-1}^L(j) + q_n(j)} \tag{16}$$

$$= (1 - \eta_n^L(j)) \cdot \Sigma_{n-1}^L(j) + \eta_n^L(j) \cdot (\mathbf{x}_n - \mu_n^L(j))(\mathbf{x}_n - \mu_n^L(j))^T .$$

The solution in Eq.(13)-(16) shows that we do not need to compute M_n^L and V_n^L, but only keep K extra scalar variables $Q_n^L(j)$ updated by Eq.(10), to compute the parameter estimates.

2.3 Case 3: O(1) Approximation

Since the weight is computed from $Q_n^L(j)$, we can in fact further eliminate the need to store $Q_n^L(j)$. However, we must consider the initialization phase when less than L samples have been processed. If the relation in Eq.(6) holds for all cases, we can replace $Q_n^L(j)$ in the learning rate Eq.(15) by $w_n^L(j)$.

Let $c(n)$ be the size of applicable window at time n, and let $\overline{Q}_n^{c(n)} = \sum_{k=1}^{K} Q_n^{c(n)}(k)$ be the cumulative posterior probability for up to L samples summed over all components. Since $\sum_{k=1}^{K} q_i(k) = 1$ for all i, then

$$\overline{Q}_n^{c(n)} = \sum_{k=1}^{K} Q_n^{c(n)}(k) = \sum_{k=1}^{K} \left[\sum_{i=n-c(n)+1}^{n} q_i(k) \right] = \sum_{i=n-c(n)+1}^{n} 1 = c(n). \tag{17}$$

Therefore, the weight computed by Eq.(6) can be expressed as for all n

$$w_n^{c(n)}(j) = \frac{Q_n^{c(n)}(j)}{\sum_{k=1}^{K} Q_n^{c(n)}(k)} = \frac{Q_n^{c(n)}(j)}{c(n)}. \tag{18}$$

We also need to verify the recursive equation in Eq.(13) for all n. Since $c(n)=L$ for $n > L$, it is trivial to replace L by $c(n)$ to verify Eq.(10) and (13). For $n \leq L$, $c(n)=n$, so $c(n)=c(n-1)+1$. Since there is no old data to be discounted, $q_{n-L}(j) = 0$, and $Q_n^{c(n)}(j)$ is computed exactly by

$$Q_n^{c(n)}(j) = Q_{n-1}^{c(n-1)}(j) + q_n(j). \tag{19}$$

Consequently, we can derive the recursive weight update equation as follows.

$$\begin{aligned} w_n^{c(n)}(j) &= \frac{Q_n^{c(n)}(j)}{c(n)} = \frac{Q_{n-1}^{c(n-1)}(j) + q_n(j)}{c(n)} = \frac{c(n-1)}{c(n)} \frac{Q_{n-1}^{c(n-1)}(j)}{c(n-1)} + \frac{q_n(j)}{c(n)} \\ &= \frac{c(n)-1}{c(n)} w_{n-1}^{c(n-1)}(j) + \frac{q_n(j)}{c(n)} = \alpha_n \cdot w_{n-1}^{c(n-1)}(j) + (1-\alpha_n) \cdot q_n(j) \end{aligned} \tag{20}$$

where $\alpha_n = (c(n)-1)/c(n)$. Based on this proof that Eq.(18) and (20) holds for all n, we can replace $Q_n^L(j)$ in the learning rate equation Eq.(15) by $c(n) \cdot w_n^{c(n)}(j)$

$$\eta_n^{c(n)}(j) = \frac{q_n(j)}{c(n) \cdot w_n^{c(n)}(j)}. \tag{21}$$

By definition, $1 \leq c(n) \leq L$. Since $w_n^{c(n)}(j)$ is updated based on $q_n(j)$, $w_n^{c(n)}(j) > 0$ if $q_n(j)>0$. If $q_n(j)=0$, then $\eta_n^{c(n)}(j)$ is set to 0. It is straight-forward to replace L with $c(n)$ to derive the same updating equations for mean and variance as Eq.(14) and (16).

$$\mu_n^{c(n)}(j) = (1-\eta_n^{c(n)}(j)) \cdot \mu_{n-1}^{c(n-1)}(j) + \eta_n^{c(t-1)}(j) \cdot \mathbf{x}_n \tag{22}$$

$$\Sigma_n^{c(n)}(j) = (1-\eta_n^{c(n)}(j)) \cdot \Sigma_{n-1}^{c(n-1)}(j) + \eta_n^{c(n)}(j) \cdot (\mathbf{x}_n - \mu_n^{c(n)}(j))(\mathbf{x}_n - \mu_n^{c(n)}(j))^T \tag{23}$$

This completes the derivation for the recursive mixture update equations based on short-term sufficient statistics with the addition of a single scalar variable $c(n)$. At each iteration, the weights are first updated based on $c(n)$ using Eq.(20), then the means and variances are updated using Eq.(22) and (23) based on learning rate calculated by Eq.(21).

2.4 Remaining Issues

In the discussion above, we have not addressed the situation when $P(\mathbf{x})=0$. Since Gaussians have infinite support, this would not occur in theory and allows us to perform the above analysis. However, in practical implementations, finite supports are imposed to determine if a match is good enough and whether an assignment should be carried out. Consequently, a data point may fall outside the boundary of all Gaussians and results in $P(\mathbf{x})=0$. In this case, one of the Gaussians G_j is set to center at the data point with a large initial variance. It can be shown that the appropriate initial weight should be $1-\alpha_n$. All other weights are updated the same way followed by normalization.

The assignment of a Gaussian does not introduce a problem when the weight of the assigned Gaussian is zero, as in the case of initialization. However, since we use a finite number of Gaussians, an existing component is reassigned when no more unused Gaussians are available. The original weight associated with that Gaussian is then implicitly distributed to all Gaussians through the normalization process. This problem is inherent to any methods using a finite number of Gaussians, and is not particular to our solution. Fortunately, since reassignment occurs only when no existing Gaussians match the data point, and does not involve mean or variance updates other than the assigned Gaussian, Eq.(21) is not used.

Another issue to be considered is whether to use the newly updated mean or the old mean when updating the variance. Since the variance is computed based on the estimated mean, the degrees of freedom are reduced by one. An unbiased estimate would subtract the old mean estimate rather than the updated one. Both formulations have been used in the literature. The choice of this estimate does not affect our derivation. In practice, there is little difference between the two as long as the variance is not initialized to 0 when the biased estimation is used.

Finally, we should mention another implicit assumption made through out our analysis. That is, the calculation of the statistics for x_t, in the expectation step, such as $q_t(k)$, should depend on the most recent parameter estimates $q_t^{t+n}(k)$ where $t+n$ is the current time. Therefore, it should be re-evaluated before including in the window. However, this is inherent in the incremental approach and it is commonly assumed that $q_t^{t+n}(k)=q_t^t(k)$.

3 Comparative Analysis

In this section, we first summarize existing solutions for online adaptive mixture learning, followed by a more detailed analysis of other algorithms based on the results derived in the previous section.

3.1 Related Work

Current applications of Gaussian mixtures for temporal distribution modeling in vision systems are exemplified by two approaches. In early application of Gaussian mixtures for video processing, mixture parameters were incrementally computed from sufficient statistics [2]. The validity of such incremental algorithms has been shown to be a stochastic approximation to the batch EM algorithm and will converge to a maximum likelihood solution [7]. However, such a formulation based on sufficient statistics is cumulative over all observed samples and cannot adapt to distribution changes. An online learning algorithm for adaptive mixtures was proposed by [8] where temporal adaptation is achieved by recursive learning. The method was shown effective and is commonly used in many vision systems. Unfortunately, learning by recursive filter converges very slowly, requiring data distribution to remain stable over a longer duration for accurate modeling. The problem was observed by [4] who proposed computing parameters based on sufficient statistics in the early stage to achieve fast convergence, then switching to recursive filter type of learning afterwards. Although their method does improve initial convergence quite significantly, we have found the formulation of the recursive learning stage to be flawed and can lead to divergence. Another approach was our own earlier work [5] where we used a modified learning rate schedule to achieve fast convergence during initial learning and gradually became recursive filter learning after many samples were observed. Although the method showed good performance experimentally, the formulation is ad hoc and no theoretical support was provided. Another formulation of adaptive mixture based on a short-term distribution was proposed by [6]. Although a simplification assumption similar to ours was made on the expected posterior, parameters calculations were formulated based on exact solutions that required $O(LK)$ amount of storage, similar to the solution in Case 1, which would be prohibitive for many video applications. The recursive updates suggest another similar derivation by applying an exponentially decaying envelop on recent samples, as suggested by [3]. In that formulation, parameters are calculated from recursively updated sufficient statistics (moments), requiring $O(K)$ overhead as in Case 2 of our solution. Furthermore, there was no description of the initial stage (or after reset) where convergence has the biggest effect. It can be shown that a $1/t$-type learning in those stages would require a uniformly weighted envelop instead of exponential.

3.2 Qualitative Analysis

All online mixture learning algorithms use the same basic adaptive filtering equation where Gaussian parameters $\theta_n(j)$ are recursively updated based on new samples

$$\theta_n(j) = (1 - \eta_n(j)) \cdot \theta_{n-1}(j) + \eta_n(j) \cdot \nabla(\mathbf{x}_n; \theta_{n-1}(j)). \tag{24}$$

Algorithms differ in the choice of $\eta_n(j)$. In the approach of [2] where parameters are calculated from incrementally updated sufficient statistics, $\eta_n(j) = 1/n$. This stochastic approximation of online EM with $1/t$-type of learning rate will converge to a local maximum likelihood solution [7]. However, the estimates are cumulative and cannot adapt to distribution changes. In the work of [8], temporal adaptation is achieved using the proposed $\eta_n(j)=\alpha g(\mathbf{x};\theta_j)$. Since the conditional probability $g(\mathbf{x};\theta_j)$ is usually very small, several researchers have found it is more efficient to simply use $\eta_n(j)=\alpha$. This type of exponential decay with fixed rate is commonly employed in many mixture-based surveillance systems for temporal adaptation. The disadvantage of this approach is slow convergence. Since α is typically set to a very small value to provide stability, any change in distribution requires a long period of learning before it is reflected in the model.

These two basic approaches can be combined to provide fast, robust adaptive mixture learning by incorporating exponential decay into the incremental EM estimates or employing a $1/t$-type of learning rate the recursive filter learning. In [4], a two-stage learning process is used. In the first stage when less than L samples are observed, the parameters are computed by incremental EM. After more than L-samples are observed, a different set of learning equations based expected sufficient statistics for the L-recent window is applied. The approach is similar to ours. However, we have shown that a distinct separation of learning stages is unnecessary and derived the optimal learning rate. More importantly, we found their formulation for the second stage of learning is flawed. Their update equations for the mean and variance can be written as

$$\theta_n(j) = \alpha \cdot \theta_{n-1}(j) + (1-\alpha) \cdot \frac{q_n(j)}{w_{n+1}(j)} \cdot \nabla(\mathbf{x}_n;\theta_{n-1}(j)) . \tag{25}$$

Since the mixing coefficients do not sum to one, we find it sometime displays divergent behavior after L samples. Another similar method was proposed by [5], who used a modified $1/t$ learning rate schedule that converges towards $(1-\alpha)$ instead of 0

$$\eta_n(j) = q_n(j)(\frac{\alpha}{Q_n^L(j)} + (1-\alpha)) . \tag{26}$$

Although it showed good performance in experiments, no theoretical justification was provided. Moreover, it requires storing an extra scalar for each Gaussian $O(K)$, similar to the solution we found in Case 2.

To illustrate the differences among these approaches, consider the following example. If a Gaussian is initialized to $\mathbf{x}$ with a variance of 1, and the same data point is observed over and over again, how fast would the variance converge to 0? Fig.1 shows the value of η_n plotted over time. The $1/t$ curve for [2] and the line $(1-\alpha)$ exemplify the two basic approaches; whereas the other three curves converge towards $(1-\alpha)$ at roughly $1/t$ rate. The middle line is the solution derived in this paper. The method of [4] follows $1/t$ for the first L samples, then follows α on the first term, and the derived line for the second term. The curve for [5] is similar to our derived solution but converges at a slower rate. For $\alpha=0.999$, it takes 100 iterations for [2][4], 91 iterations for [5] and 95 iterations for Eq.(21) for σ to reach 0.01, compared to 4603 iterations for the commonly used method in [8].

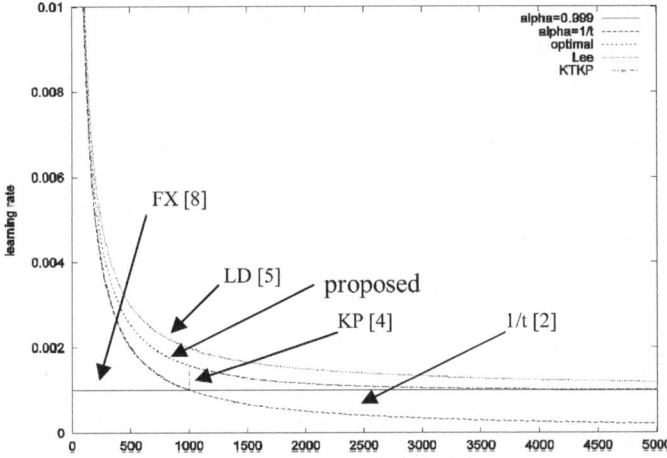

Fig. 1. A comparison of effective learning rates of several algorithms for α=0.999

4   Experimental Results

For a quantitative comparison, we evaluated various algorithms on a set of synthetically generated data with known ground truth. Each data set simulates pixel intensity
values in video over time and is generated by randomly drawing a value between 0
and 255 from a time-varying mixture distribution with known parameters. These
synthetic data models are created based on observations of real video data. Fig.2
shows some examples where the intensity value (y-axis) is plotted over time (x-axis).
There are three types of data models. Type 1 consists of a tri-modal mixture density
with one dominant cluster and two smaller clusters. This distribution is representative
of regions in a surveillance video where foreground activities are observed. Type 2
data is composed of a single migrating distribution, aimed at testing the temporal
adaptability of the models. Type 3 data consists of a bimodal distribution observed in
real videos due to monitor flickering or swaying trees.

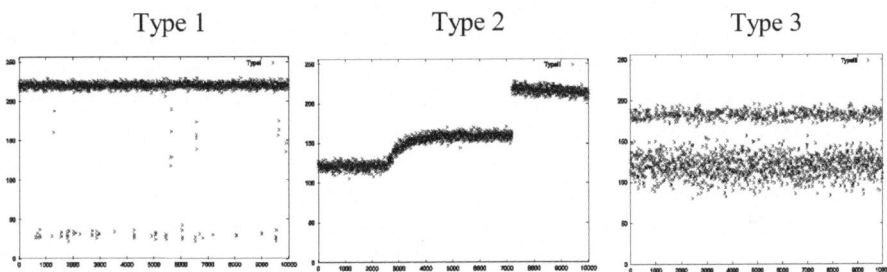

Fig. 2. Examples of synthetic data sets modeling pixel intensity values in video

Four adaptive algorithms were tested: the fixed rate method in [8] as a baseline (FX), two previously proposed methods for improving convergence [4] (KP) and [5] (LD), and the solution derived in this paper (Proposed). At every iteration, we compute the KL-divergence between the estimated model and the ground truth. The algorithms are trained for $5L$ iterations where L is the adaptive window size. To get a sense of how fast the algorithm converges and adapts, we compute the average divergence over the entire learning window. The results are summarized in Table 1. Each entry is averaged over 10 data sets.

Table 1. Performance comparison for different adaptive learning algorithms. Each entry in the table is the KL-divergence between the estimated model to the distribution ground truth, averaged over 10 data sets on $5L$ iterations, where $L=1/(1-\alpha)$ is the adaptive window size

KL-Divergence	FX	KP	LD	Proposed
(a) Type1,α=0.999,K=3	0.006866	0.007563	0.000148	0.000148
(b) Type1,α=0.999,K=5	0.006866	0.008976	0.000123	0.000124
(c) Type1,α=0.99,K=3	0.006453	0.003868	0.000644	0.000725
(d) Type1,α=0.99,K=5	0.006453	0.004885	0.000644	0.000726
(e) Type2,α=0.99,K=3	0.010331	0.000520	0.000471	0.000520
(f) Type2,α=0.999,K=3	0.018361	0.015017	0.003878	0.004363
(g) Type3,α=0.99,K=3	0.008844	0.010889	0.001666	0.001668
(h) Type3,α=0.999,K=3	0.008494	0.010102	0.001413	0.001426

The experimental results are consistent with our analysis in the previous section. The commonly used fixed rate method (FX) is orders of magnitude slower in convergence than any other adaptive methods, resulting in proportionally larger average divergence. The KP method performed well in the initial learning stage. However, learning performance in the second stage is sensitive to data characteristics and parameter settings. For example, its performance in trial (e) is comparable to the LD and proposed method. However, a similar run with different parameters in (f) is much worse relative to others because of several diverging cases. The unstable nature of the algorithm can be seen by the inconsistent results shown across different tests, as predicted in our analysis. The LD, which was our own method developed earlier, and the proposed methods performed equally well across all test sets. The slight advantage for the LD method can be attributed to a slower decreasing learning rate (as shown in Fig.1). However, it requires additional storage of $O(K)$ variables. In addition, without any theoretical backing, it is difficult to quantify its expected behavior.

We also tested the algorithms on real surveillance videos, as shown in Fig. 3. An adaptive mixture is used to model the color distribution at a pixel in the video over time. We used a separate mixture for each pixel with 3 Gaussian components in the YUV space and a diagonal covariance matrix, and α=0.995. We compute the likelihood of each new frame under the previous estimates and plotted over time, shown on the right hand side. The observation again is consistent with our pretion. Model trained under the FX method improved very slowly compared to others.

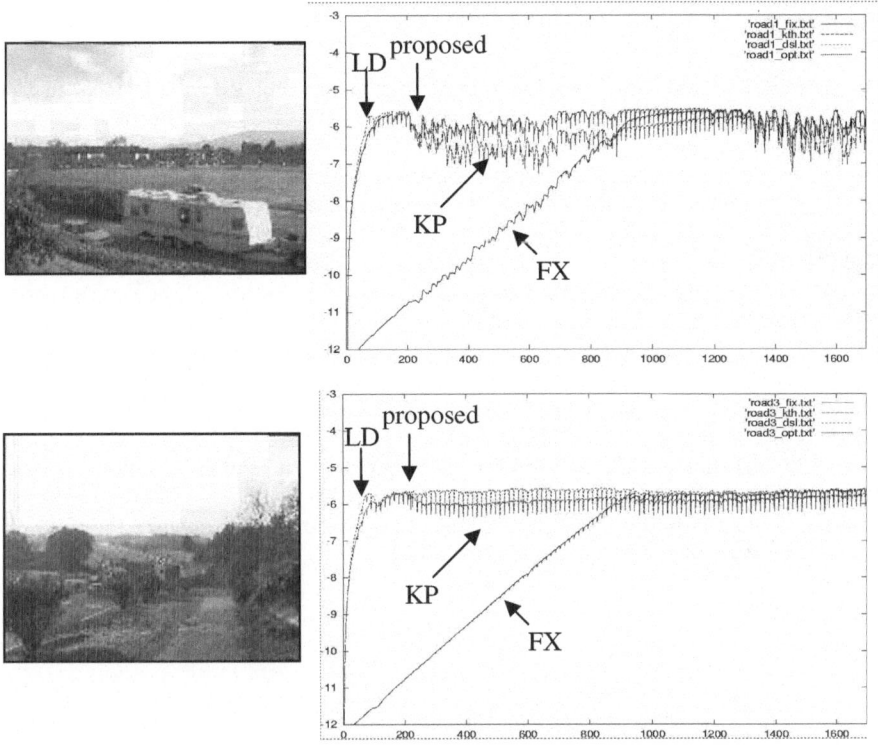

Fig. 3. Testing on traffic monitor video. The log likelihood curve of video predicted by various algorithms are compared

The KD method performed quite well except for frames between 300 and 800 when moving traffic come into view. Similar to the previous tests on Type 1 data, the algorithm diverged at several locations in the video and lowered the overall frame likelihood. The LD and proposed method were again on par with each other, with the LD method converging very slightly ahead at the beginning because of a slower decreasing learning rate. The exact same observations can be made on the second sequence as well.

5 Conclusion

We presented an online EM learning algorithm for training adaptive Gaussian mixtures. We derived a set of recursive parameter update equations based on short term sufficient statistics that can be computed without additional storage of auxiliary variables. To our knowledge, this is the first such derived formulation of its kind. The proposed solution was evaluated against existing algorithms and showed superior efficiency and robustness on large simulations as well as real video data.

References

1. Dempster, A.P., Laird, N.M. and Rubin, D.B.: Maximum likelihood from incomplete data via the EM algorithm. J. of the Royal Statistical Society B, vol. 39 (1977) 1-38.
2. Friedman, N. and Russell, S.: Image segmentation in video sequences: a probabilistic approach. In: Proc. 13th Conf. Uncertainty in Artificial Intelligence, (1997).
3. Jepson, A.D., Fleet, D.J., El-Maraghi, T.: Robust online appearance models for visual tracking. IEEE Trans. on PAMI Vol. 25, No.10, (2003) 1296-1311.
4. KaewTraKulPong, P. and Bowden, R.: An improved adaptive background mixture model for real-time tracking with shadow detection. In: Proc. of 2nd Euro. workshop on Advanced Video Based Surveillance Systems (2001).
5. Lee, D.S.: Improved adaptive mixture learning for robust video background modeling. In: Proc. of IAPR Workshop on Machine Vision for Applications (2002) 443-446.
6. McKenna, S.J., Raja Y. and Gong, S.: Object tracking using adaptive colour mixture models. In: Proc. of ACCV, vol.1, (1998) 615-622.
7. Neal, R.M. and Hinton, G.E.: A view of the EM algorithm that justifies incremental, sparse, and other variants. In: Learning in Graphical Models (1998).
8. Stauffer, C. and Grimson, W.E.L.: Adaptive background mixture models for real-time tracking, In: Proc. of CVPR, vol.2, (1999) 246-252.

Novelty Detection in Image Sequences with Dynamic Background

Fredrik Kahl[1], Richard Hartley[1][2], and Volker Hilsenstein[3]

[1] Dept. of Systems Engineering, Australian National University, Australia
[2] National ICT, Australia
[3] Institute for Environmental Physics, University of Heidelberg, Germany

Abstract. We propose a new scheme for novelty detection in image sequences capable of handling non-stationary background scenarious, such as waving trees, rain and snow. Novelty detection is the problem of classifying new observations from previous samples, as either novel or belonging to the background class. An adaptive background model, based on a linear PCA model in combination with local, spatial transformations, allows us to robustly model a variety of appearances. An incremental PCA algorithm is used, resulting in a fast and efficient detection algorithm. The system has been successfully applied to a number of different (outdoor) scenarious and compared to other approaches.

1 Introduction

Given a stationary camera taking images of an outdoor scene, the problem considered in this paper is to detect novel events in the sequence of images. A *novel* event can loosely be defined as an event that cannot be inferred from previous images. Many times, it comes down to separating foreground objects from the (possibly changing) background.

The detection of novel events is often the first stage in many visual surveillance systems. Typically, background subtraction is a method employed to alert when new events occur in the sequence. The information provided by this low-level system is an important cue and it can be used for more high-level tasks, such as motion tracking, recognition of events, etc. See [2] for a recent review of the current state-of-the-art in video surveillance.

The requirements for such a novelty detection system are that it must work in (near) real-time, the background should be adaptive and be able to deal with illumination changes, and preferably work with both grey-scale and colour imagery. A common assumption is that the background is static, which makes things simple on one hand, but limits the applicability on the other hand. In this paper we try to relax this assumption and to cope with non-stationary data in the background by introducing a more flexible background model. It is a linear PCA model, which makes it possible to model different appearances. In order to allow for local deformations, the image plane is partitioned into a set of (possibly overlapping) regions which may move locally around its origin. The model is adaptive, and to speed up computations an incremental PCA algorithm is developed based an algorithm called PowerFactorization ([4]).

D. Comaniciu et al. (Eds.): SMVP 2004, LNCS 3247, pp. 117–128, 2004.

1.1   Related Work

There is a huge literature on background subtraction/modelling for static (or slowly changing) background scenes. For example, recursive updates of the background model using an adaptive filter was used in [11] and a Kalman filter in [8]. In [7], a statistical background model was presented focusing on handling illumination changes. A breakthrough in background modelling was made by Stauffer and Grimson in [10] who used a mixture of Gaussians to model multiple hypotheses. To reduce the complexity of the model, each pixel was treated independently. A similar approach was taken by Elgammal et al. [3], but instead kernel methods were applied to obtain non-parametric estimates of the probability distributions of the pixels. Again, no interdependencies between pixels were assumed. Even though these approaches have proven to be successful for many scenarios, they do not handle very much dynamic motion in the background.

In a recent paper by Monnet et al. [5], a background model that explicitly tries to model dynamic textures is presented. The dynamic model is based on the work in [9]. Our model has many resemblances with their model as they employ linear PCA as well. However, they pursue another line of thought. The coefficients of the linear model are fed into an autoregressive dynamical system and then they use the system as a prediction/detection mechanism for novel behaviour. Their model is adaptively updated, but it is unclear how they handle older model coefficients. When the PCA basis is updated, these coefficients become obsolete.

2   Novelty Detection

Before delving into our approach to the problem, we will give a more formal problem formulation. Novelty detection is concerned with first, estimating the unknown parameters (or latent variables) of a statistical model from a set of observations and then, for subsequent observations, deciding whether they should be regarded as novel or not. In a statistical setting, this can be formulated as follows. Given a set of observations $\{x_1, \ldots, x_t\}$, we want to estimate the latent variables θ for some probability distribution $p(x|\theta)$, for example, by applying the Maximum Likelihood method. Then, a new observation x_{t+1} can be considered as *novel* if $p(x_{t+1}) < p_0$ for some threshold p_0.

In our setting, the observables are images $\{I_i\}_{i=1}^t$, where $I_i : \Omega \to \mathbf{R}$ and Ω denotes the image plane . Suppose we have a model $\mathcal{M}$ describing this set (or some representation of it), then novel detection becomes:

$$\text{Is } I_{t+1} \in \mathcal{M} \ ?$$

Or, in words, is I_{t+1} close to $\mathcal{M}$ in some suitable metric?

In addition, we want our statistical model to evolve over time, that is, to adapt to new data, such that only sudden changes are detected. Hence, given $\{I_i\}_{i=1}^{t+1}$ and the current model $\mathcal{M}_t$, it should be easy to infer an updated model $\mathcal{M}_{t+1}$. The influence of older observations should generally be lower than for new ones - the system has only limited memory.

If the resulting novelty detection scheme is to be practically useful, it needs to be able to process the data in real-time. Hence, there is a compromise between the level of complexity of the model and the computational speed. Therefore, it is essential to consider solutions that are able to update the model parameters incrementally.

3 Background Model

This section introduces the statistical background model $\mathcal{M}$ for our novelty detection scheme. Again, remember that the set of observations consists of 2D images of a scene and often the background is non-stationary. For example, trees sway or it could be raining. A sequence of images contains a lot of data, and hence the complexity of the model and the methods have to be restricted.

Linear models are a good compromise between low complexity and reasonable approximation of the observed data – witness the success of Active Appearance Models [1]. Suppose an image I_t at time t can be modelled by an affine function plus a noise term:

$$I_t = I_{mean} + \sum_{i=1}^{r} \lambda_{it} \Phi_i + \epsilon, \tag{1}$$

where I_{mean} can be thought of as the "mean image", the Φ_i are some vectors describing modes of possible variation, and the λ_{it} are some time varying scalars. The noise ϵ is supposed to be zero-mean and normally distributed[1].

The above model for $\mathcal{M}$ maintains interdependencies between pixels in a natural way, and it is capable of modelling a variety of different (global) appearances [1]. However, it does not account for local perturbations well (such as swaying trees) if only a few modes of variation are incorporated into the model.

In order to achieve more locality, we partition the image plane Ω into a set of (possibly overlapping) regions Ω_i such that $\Omega = \cup_i \Omega_i$. Then, we can apply the model in (1) to each region independently. In addition, we will allow for small spatial transformations of each region. Let $T : \mathbf{R}^2 \rightarrow \mathbf{R}^2$ denote a spatial transformation, then an image region Ω_i can be modelled by

$$I_t(\mathbf{x}) = I_{mean}(T(\mathbf{x})) + \sum_{i=1}^{r} \lambda_{it} \Phi_i(T(\mathbf{x})) + \epsilon. \tag{2}$$

We will only allow for small translations, i.e., $T(\mathbf{x}) = \mathbf{x} + \mathbf{\Delta}x$ where $||\mathbf{\Delta}x||$ is small, on the order of a few pixels. Otherwise, the region might start tracking moving foreground objects. For more details, see Section 5 on experimental results.

In principle, there are many ways to partition the image plane. For simplicitly, we have partitioned the image into blocks. The size of these blocks is a crucial parameter. If the blocks are too large, then the model is less capable of capturing

---

[1] It is possible to assume a prior on λ_{it} as well, but in our experiments that assumption has had no noticeable effect.

local dynamics. And too small blocks will tend to, in the limit, to a pixel-based model. The issue is further discussed in the experimental section.

4   Estimation of Model Parameters

There are essentially three sets of parameters that need to be estimated in the algorithm: the mean, the linear basis and the spatial transformations. The computation of the first two are described in the next section. Then follows an outline of the complete algorithm (including the estimation of the third set of parameters).

4.1   PowerFactorization

At the heart of our background subtraction algorithm is the extraction of a Principal Component Analysis (PCA) basis for the image blocks. Consider a particular region Ω_i in the image. The pixels in this block at time t may be represented by a vector $\mathbf{v}_{it}$. (Note that the pixels in $\mathbf{v}_{it}$ need not come from precisely the same position in the image at each time, since small motions of the blocks are allowed for.) For instance, in the experiments described later, each block is a block of size 17×23, so $\mathbf{v}_t$ is a vector of dimension $D = 17 \times 23 = 391$. For simplicity, we drop the subscript i, and bear in mind that the succeeding discussion refers to each block Ω_i in the image.

In order to keep a finite memory of the previous appearances of the block, we consider some number N of previous states of the block, represented by N vectors $\mathbf{v}_{t-N+1}, \ldots, \mathbf{v}_t$. In the experiments described below, $N = 50$. We denote this set of vector by $V_t = \{\mathbf{v}_{t-N+1}, \ldots, \mathbf{v}_t\}$. Our purpose is to carry out PCA on these vectors at each time t to find a small number r of such vectors that most nearly span the space containing the complete sequence $\mathbf{v}_{t-N+1}, \ldots, \mathbf{v}_t$.

At each time t, we may compute the vector $\bar{\mathbf{v}}_t$, which is the mean of all the vectors in V_t. Subtracting this mean from each $\mathbf{v}_t$, and numbering appropriately, we obtain a zero-mean set of vectors $W_t = \{\mathbf{w}_{t1}, \mathbf{w}_{t2}, \ldots \mathbf{w}_{tN}\}$, where $\mathbf{w}_{ti} = \mathbf{v}_{t-i+1} - \bar{\mathbf{v}}_t$.

To carry out PCA, on the set of vectors W_t, we form a matrix $\mathtt{M}$, the columns of which are the vectors in W_t. Thus, $\mathtt{M}$ has dimension $D \times N = 391 \times 50$ in the experiments. We wish to represent the column space of $\mathtt{M}$ by a small number r of vectors. In the experiments, $r = 4$. Equivalently, we wish to find the matrix $\widehat{\mathtt{M}}$ of rank r that is closest to $\mathtt{M}$ in an appropriate norm – the Frobenius norm. A common way that this may be done is by using the Singular Value Decomposition (SVD). Let $\mathtt{M} = \mathtt{U}\mathtt{D}\mathtt{V}^\top$, where the diagonal entries of $\mathtt{D}$ (the singular values) are in descending order. Then the first r columns of $\mathtt{U}$ form an orthonormal basis for the dimension-r subspace that we require, our PCA basis. Equivalently, the closest rank-r matrix to $\mathtt{M}$ is the matrix $\widehat{\mathtt{M}} = \mathtt{U}\mathtt{D}^{(r)}\mathtt{V}^\top$, where $\mathtt{D}^{(r)}$ represents the (diagonal) matrix formed from $\mathtt{D}$ by setting all but the first r diagonal entries to zero.

The SVD method gives an exact solution. Unfortunately, the computational cost of this algorithm is quite high, particularly since we need to carry it out for each block Ω_i in each frame. Instead, we make use of the fact that the PCA basis for a given block will not change very much from one time instant to the next. Consequently, an iterative update method is more suitable. We use the method of PowerFactorization presented in [4] to accomplish this. It should be noted that another (similar though not identical) method for iterative low-rank approximation has been suggested by [6].

We are carrying out PCA on a sliding window of N vectors derived from each block Ω_i at successive times, indexed by t. It is clear that the mean vector $\bar{\mathbf{v}}_t$ can easily be computed recursively with minimal effort at each time step. Similarly a matrix containing (as columns) the last N vectors is easily maintained by writing each new vector $\mathbf{v}_t$ over the top of the vector $\mathbf{v}_{t-N}$ that it is replacing. Then, the mean $\bar{\mathbf{v}}_t$ can be subtracted from each column to obtain the matrix M of dimension $D \times N$.

In the PowerFactorization algorithm, we estimate two matrices A of dimension $D \times r$ and B of dimension $N \times r$, such that $\widehat{\mathsf{M}} = \mathsf{AB}^\top$ is the closest rank-r approximation to M. In addition A has orthonormal columns, which therefore form the vectors of the desired PCA basis. This is done by a simple iterative procedure as will be described soon. At each time instant t, the matrices A and B are computed iteratively, and for simplicity, we will omit the index t representing the time instant (or frame number). Subscripts represent the iteration number within the iterative procedure at a given time instant.

The iteration consists of three steps, starting from an initial value for the matrix A_0.

1. Define $\mathsf{B}_k = \mathsf{M}^\top \mathsf{A}_k$.
2. Define $\mathsf{A}_{k+1} = \mathsf{MB}_k$.
3. Apply the Gram-Schmidt algorithm (otherwise known as QR-factorization) to orthonormalize the columns of A_{k+1}.

The product $\mathsf{A}_k\mathsf{B}_k^\top$ is guaranteed ([4]) to converge linearly (even from a random starting point) to the closest rank-r matrix $\widehat{\mathsf{M}}$ to M. In particular, for some constant K,

$$\|\widehat{\mathsf{M}} - \mathsf{A}_k\mathsf{B}_k^\top\| \leq K(s_{r+1}/s_r)^k$$

where s_i is the i-th greatest singular value of M.

It remains to explain how to choose an initial value for A_0. Under the assumption that the PCA basis does not change much from one time instant to the next, the best strategy is to start from where we left off at the previous time instant. In particular, since A contains the PCA basis, we start with A_0 equal to the final matrix A_{k+1} from the previous time instant. At the very start, when no previous value for A is available, we may begin with a random value of A_0, and the algorithm will boot-strap itself.

Note that convergence of this algorithm is quickest when the ratio s_{r+1}/s_r is small – that is, the matrix is close to having rank r. This may not be the case in the present problem, but it does not cause significant difficulty, since it is not essential to have the absolute best PCA basis. In practice, a small number (we use 3) of iterations at each time step are sufficient. In our tests, this was enough to ensure that the vectors of the orthonormal PCA basis found in this way were within 1% of a true spanning set for the subspace found using the exact SVD algorithm.

4.2   Outline of the Algorithm

As the residuals between the model and the measurements are assumed to be Gaussian, cf. (2), it implies that the Frobenius norm of the residuals is the statistically correct measure to use for deciding whether a new sample is novel or not. For a given position, the coffecients λ_{it}, $i = 1, \ldots, r$ are computed by a projecting onto the basis spanned by Φ_i (which is done by a scalar product as the basis vectors are orthogonal). The residual error is computed by

$$\epsilon^2(I_t) = ||I_t - I_{mean}||_F^2 - \sum_{i=1}^{r} \lambda_{it}^2.$$

For each region in the image, the following steps are performed at time t.

1. Let $\mathbf{x}_{t-1}$ be the position in the previous image of the region. Compute the residual error ϵ at this position.
2. If $\epsilon < p_{low}$, then set $\mathbf{x}_t = \mathbf{x}_{t-1}$ and go to 5.
3. For each integer position in a neighbourhood of the previous position, compute the residual error and set $\mathbf{x}_t$ to the one with lowest error.
4. If the residual error ϵ at $\mathbf{x}_t$ is greater than p_0, then declare the region as novel.
5. Update the mean and the PCA basis incrementally using PowerFactorization.

The test in the second step above is done to speed up the algorithm. Most regions in the image do not move and hence it is not necessary to check for other translations if the reconstruction error is low for no movement. (In the experiments, $p_{low} = p_0/3$.) The neighbourhood of $\mathbf{x}_{t-1}$ is defined to include all points within a circle of the region's original position. The radius of the circle is set to 10 pixels and for computational reasons, we assume that a region moves at maximum two pixels per frame. The threshold p_0 is based on the estimated standard deviation of the $N = 50$ previous residual errors (which can be computed incrementally).

5   Experimental Results

We have tested the proposed algorithm on a number of sequences with promising results. Some typical behaviour of the algorithm on a representative selection of sample sequences are presented in this section. The algorithm has also been compared to the approach of Elgammal et al. [3] and to the Stauffer-Grimson approach [10], which can be considered to be the state-of-the-art in terms of real-time background modelling. Both of these two alternative approaches are pixel-based (see Section 1.1) and can handle multi-modal probability pixel distributions. The main difference between them is in the way the probabilitiy distributions are estimated, but the performance is similar.

The sequences presented below consist of frames with 240×320 grey-scale pixels. It is straightforward to apply the algorithm on colour images as well, though with increased computational requirements. The image has been divided into rectangle regions of 17×23 pixels, such that it is covered by a total of 15×15 rectangles. Each region is allowed to translate up to 10 pixels from its origin (but at most 2 pixels per frame for computational reasons). In the model, the number of principal components is set to 4 for each rectangle. The model is updated using the 50 previous frames (those that are not classified as novel) using the incremental PowerFactorization algorithm. The number of iterations in the innerloop of PowerFactorization is set to 3. We have experimentally validated that even with so few iterations, we approximate the true linear subspace within one percent in Frobenius norm. Our matlab implementation can perform novelty detection with 3 frames per second on a Pentium 4, 2.9 GHz.

In the left of Figure 1, one image of the forest sequence of Elgammal[2] is shown. It consists of a person walking in the woods. Partial occlusions occur due to branches and the trees sway slightly - the wind is quite moderate. Just looking at a single frame, it is hard to detect the person, however, the algorithm has no problem in tracking the person. The detected squares are marked in the middle of Figure 1. The sequence has also been tested with the algorithm of Elgammal [3] and here the result is even better. As this approach is pixel-based, one gets a more precise localization of the foreground object compared to a region-based approach.

Fig. 1. Left: An example frame in the Elgammal forest sequence. **Middle**: Output of our algorithm. Rectangles detected as novel are marked. **Right**: Thresholded result of Elgammal [3]

The next sequence shows a traffic intersection (also from Elgammal), Figure 2. The difficulty in this sequence is due to the heavy rain, which may be hard to spot in a single frame. The results of both our and Elgammal's algorithms are given in the same figure. As can be seen, the two algorithms perform well, though, the techniques for handling the distortions caused by the rain are quite different. In the first case, the PCA model finds a suitable subspace for each region, while in the latter case, the effects are modelled by a non-parametric estimate of the pdf on a pixel basis.

Fig. 2. Left: One frame in the Elgammal rain sequence, followed by the results of our algorithm (**middle**) and Elgammal (**right**)

The next sequence is similar to the previous one, except that the camera is not stable - it shakes a bit and as a result the image jumps up and down a few pixels in the sequence. The output of our algorithm is presented in Figure 3 and as can be seen, the spatial transformations (i.e., the translations) are able to compensate well for the shaking image. In the left of Figure 4 the result of Elgammal's standard pixel-based algorithm is shown - a lot of pixels are classified as novel. Elgammal et al. have also developed a motion-compensated version of their algorithm, which tests if a given pixel can be explained by neighbouring pixel distributions (see right of Figure 4). Now, the shakiness is eliminated from the output.

So far, we have only demonstrated that our algorithm works fine where alternative approaches may also work well. The next sequence is somewhat harder. The scene depicts a road with a swaying tree and a lot of cast shadows all over the image. Intermittent gusts of wind cause vivid motion of the tree's branches and the corresponding shadows to move vigorously. Three frames in the sequence are shown in Figure 5 and the corresponding detection results of our scheme in Figure 6 and Stauffer-Grimson in Figure 7.

We have tested this sequence with the Stauffer-Grimson algorithm[3] with up to 8 Gaussians in the mixture, but without great success. The pixelwise distributions are not able to capture the non-stationary texture well. As a comparison,

---

[3] Thanks to D. Magee, University of Leeds, for making the algorithm publicly available.

Fig. 3. Left: One frame in the shaking camera sequence. **Middle**: The output of our algorithm

Fig. 4. The result of Elgammal witout motion-compensation (**left**) and with motion-compensation (**right**)

Fig. 5. Three frames of the Daley road sequence. Notice, in the right image a cyclist can be spotted

we have graphed the percentage of pixels that are classified as novel with their algorithm and similarly with ours (Figure 8). Notice that even though there are high peaks where foreground objects appear, there are also peaks at other times where there is no actual novelty. The situation looks much better in the second case, even though a few blocks are occasionally misclassified.

Fig. 6. The detection results of our algorithm for the images in Figure 5

Fig. 7. The detection results of Stauffer-Grimson algorithm for the images in Figure 5

5.1   Limitations

Testing the algorithm on a number of sequences, we have found some drawbacks. One parameter that is crucial is the size of regions in the images.

- We have noticed at some occasions that if the novel foreground object is relatively small compared to the region size, the region may not flag it as novel. Instead, the small foreground object is temporarily merged into the background model.
- Compared to a pixel-based methods, the detection boundary of the foreground objects is less precise.
- Slowly moving foreground objects may adaptively be merged into the background model. In the current setting, this hardly ever happens, as objects would need to move quite slowly.

A limitation of the current paper is of course that it lacks a full experimental comparison to the work of Monnet et al. [5]. It seems likely that their model is more suitable for truly, repetitive patterns like ocean waves, but our approach would perform better on scenes with local, spatial dynamics like waving trees. The implementation of Elgammal's algorithm was neither available, therefore only a comparison of the demonstration scenes provided on his web page was possible.

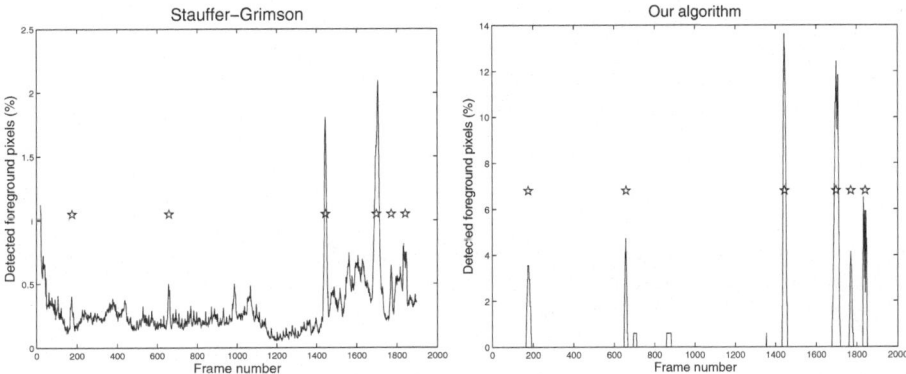

Fig. 8. Results of Stauffer-Grimson (**left**) and our algorithm (**right**) on Daley Road sequence. The stars indicate where actual cars or cyclists pass by

6   Conclusions

In this paper, we have developed a new scheme for novelty detection in image sequences, capable of handling non-stationary background scenarious. The main advantages of the system are its simplicity, the powerful combination of a PCA model with local, spatial transformations and a fast, incremental PCA through PowerFactorization.

References

1. T. F. Cootes, G. J. Edwards, and Taylor C. J. Active appearance models. *IEEE Trans. Pattern Analysis and Machine Intelligence*, 23(6):681–685, 2001.
2. A. Dick and M.J. Brooks. Issues in automated visual surveillance. In *International Conference on Digital Image Computing: Techniques and Applications*, Sydney, Australia, 2003.
3. A. Elgammal, D. Harwood, and L. S. Davis. Non-parametric model for background subtraction. In *European Conf. Computer Vision*, volume II, pages 751–767, Dublin, Ireland, 2000.
4. R. Hartley and F. Schaffalitzky. Powerfactorization : Affine reconstruction with missing or uncertain data. In *Proc. IEEE Conference on Computer Vision and Pattern Recognition, Washington DC*, 2004 (submitted).
5. A. Monnet, A. Mittal, N. Paragios, and V. Ramesh. Background modeling and subtraction of dynamic scenes. In *Int. Conf. Computer Vision*, pages 1305–1312, Nice, France, 2003.
6. T. Morita and T Kanade. A sequential factorization method for recovering shape and motion from image streams. *IEEE Transactions on Pattern Analysis and Machine Intelligence*, 19(8):858–867, 1997.
7. N. Ohta. A statistical approach to background substraction for surveillance systems. In *Int. Conf. Computer Vision*, volume II, pages 481–486, Vancouver, Canada, 2001.

8. C. Ridder, O. Munkelt, and H. Kirchner. Adaptive background estimation and foreground detection using kalman filtering. In *Int. Conf. on Recent Advances in Mechatronics*, pages 193–199, 1995.

9. S. Soatto, G. Doretto, and Y.N. Wu. Dynamic textures. In *Int. Conf. Computer Vision*, volume II, pages 439–446, Vancouver, Canada, 2001.

10. C. Stauffer and W.E.L. Grimson. Adaptive background mixture models for real-time tracking. In *Conf. Computer Vision and Pattern Recognition*, volume II, pages 246–252, Fort Collins, USA, 1999.

11. C.R. Wren, A. Azarbayejani, T. Darrell, and A. P. Pentland. Pfinder: real-time tracking of the human body. *IEEE Trans. Pattern Analysis and Machine Intelligence*, 19(7):780–785, 1997.

A Framework for Foreground Detection in Complex Environments

Junxian Wang, How-Lung Eng, Alvin H. Kam, and Wei-Yun Yau

Institute for Infocomm Research,
21 Heng Mui Keng Terrace, Singapore 119613
junxian@i2r.a-star.edu.sg

Abstract. In this paper, a framework is proposed for the foreground detection in various complex environments. This method integrates the detection and tracking procedures into a unified probability framework by considering the spatial, spectral and temporal information of pixels to model different complex backgrounds. Firstly, a Bayesian framework, which combines the prior distribution of the pixel's features and the likelihood probability with a homogeneous region-based background model, is introduced to classify pixels into foreground and background. Secondly, an updating scheme, which includes an on-line learning process of the prior probability and background model updating, is employed to guarantee the accuracy of accumulated statistical knowledge of pixels over time when environmental conditions are changed. By minimizing the difference between the priori and the posterior distribution of pixels within a short-term temporal frame buffer, a recursive on-line prior probability learning scheme enables the system to rapidly converge to the new equilibrium condition in response to the gradual environmental changes. This framework is demonstrated in a variety of environments including swimming pools, shopping malls, office and campuses. Compared with existing methods, this proposed methodology is more robust and efficient.

1 Introduction

Within the last ten years, there has been a remarkable increased interest in understanding and developing automated surveillance systems, which provide intelligent recognition of events or human behavior in an unsupervised way. Generally, there are three fundamental components in building such a system: change detection, tracking and event inference. In a real-time surveillance system, the effective implementation of tracking and sophisticated inference depend heavily on the reliability of the change detection process. In order to improve the performance of change detection, various change detection algorithms have been proposed, such as background subtraction, frame differencing and optical flow. These existing methods separately or jointly exploited the spatial (e.g., local structure and gradient), and temporal features (e.g., correlation w.r.t. a reference frame and inter-frame correlation) of video sequences for foreground

D. Comaniciu et al. (Eds.): SMVP 2004, LNCS 3247, pp. 129–140, 2004.

detection. Many existing methods [1], utilized spatial features to adapt to environmental changes. However, these spatial-based approaches overlooked the fact that the frames are continuous in video sequences. Thus, it is difficult to describe the appearance of pixels in dynamic background. Usually, the performance of the change detection process can be improved by jointly considering spatial and temporal features to represent the complex background scene containing both stationary scene and non-stationary changes (temporal change). Li [2] exploited the color co-occurrence (dynamic features), colors and gradient of local structure (static feature) for the change detection to distinguish the dynamic and static property of pixels, respectively. However, their change detection and tracking process are independent on each other. As a result, the only information can be used at one moment and there is the loss of tracking knowledge's effect to change detection. In order to provide the inter-feedback between the tracking knowledge and the change detection, Ragini [3] integrates the information of face probabilities provided by the detector and temporal information provided by the tracker for detecting and tracking multiple faces in a video sequence. It used local histograms of wavelet coefficients represented with respect to a coordinate frame fixed to the object. The probabilities of detection are propagated over time by using a zero order prediction model on the detection parameters. Ragini's method helps to stabilize detection over time. However, this approach needs to overcome the high computation and time complexity to compute coefficients of wavelet transformation for real-time application.

In order to meet the real-time requirement and improve the performance of change detection based on the inter-feedback between the tracking knowledge and change detection, we propose a novel framework to integrate the detection and tracking procedures into a unified probability framework by incorporating spatial, spectral and temporal information of pixels. The framework includes:

1. A pixel-based Bayesian decision approach, which incorporates the prior knowledge of pixels and likelihood probability of the pixels, is introduced. The change detection based on this approach is determined by the accumulated statistical knowledge of pixels over time.
2. An on-line recursive learning process, which minimizes the difference of the prior distribution and posterior distribution of pixels over a short frame buffer, is introduced. This process guarantees the accuracy of the accumulated prior distribution of pixels for change detection.
3. The representation of background model is developed based on neighborhood information of a pixel, homogeneous information in a square block and temporal information of adjacent frames. This background model considers both the dynamical property of background over a short duration and static property of background over a long duration. Thus, the foreground false alarm is efficiently suppressed.

In order to evaluate the performance of the proposed framework, extensive experiments have been designed and performed at the swimming pool, the shopping mall, the office and the campus. The comparison of the detection results between the proposed framework and other existing methods, indicates that the proposed approach is more robust and effective.

2   Bayesian Framework for Object Detection and Tracking

In this work, we formulated the problem of foreground detection and tracking under various complex environments into a Bayesian framework. It includes four phases — learning phase, detection phase and tracking phase and updating phase (shown in Figure 1).

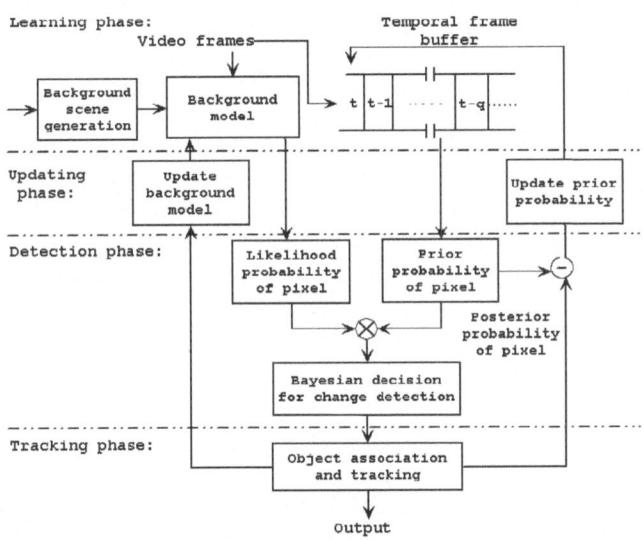

Fig. 1. The block diagram of the proposed detection and tracking algorithm

Firstly, a background model is built by incorporating spatial information of pixels and homogeneous information in a square block. Then, a Bayesian decision classified pixels in the current frame into foreground and background by jointly considering the likelihood probability of pixels with the background model and the prior distribution of pixels in adjacent frames. In which, a high order Hidden Markov model (HMM) with mixture transition distribution is introduced to estimate similar features from previous frames. Then, objects are formed by labeling foreground pixels. In order to guarantee the accuracy of the accumulated prior distribution of pixels for change detection, an on-line learning scheme is proposed to minimize the difference between the priori distribution and posterior distribution of pixels in a short frame buffer. Meanwhile, the background model was updated to adapt to dramatic environmental changes. The following subsections will describe the proposed framework and the associated components in more detail.

2.1   Bayesian Foreground Detection Scheme

Let's consider a sequence of background frames with size $H \times L$ from tth frame to $t-q$th frame, and divide each frame into $N_H \times N_L$ non-overlapping square blocks with size of $s \times s$ for each, where $N_H = H/s$ and $N_L = L/s$. Let $X_{a,b} = \{x_{i+k,j+m}^{t-1}, \cdots, x_{i+k,j+m}^{t-q}\}$ be pixels collected from square blocks of the background frames with position index (a,b), where $1 < a < N_H$ and $1 < b < N_L$. $x_{i+k,j+m}^t$ are eight neighboring pixels around $x_{i,j}^t$, $a*s < i+k < (a+1)*s$, $b*s < j+m < (b+1)*s$. A set of homogeneous regions, $\{R_{a,b}^1, \cdots, R_{a,b}^c\}$, of the background is formed by clustering every $X_{a,b}$ [4].

Therefore, the problem of foreground detection can be translated as classifying $x_{i,j}^t$ pixel into a symbol $\boldsymbol{\Theta}_{a,b} = \{\boldsymbol{\theta}_{a,b}^B, \boldsymbol{\theta}_{a,b}^F\}$, where $\boldsymbol{\theta}_{a,b}^B$ and $\boldsymbol{\theta}_{a,b}^F$ denote the background and foreground model in square blocks with position index (a,b), respectively. This framework for foreground detection is formulated as follows:

$$P(\boldsymbol{\Theta}_{a,b}|x_{i,j}^t, \cdots, x_{i,j}^{t-q}) \propto P(\boldsymbol{\Theta}_{a,b}|x_{i,j}^{t-1}, \cdots, x_{i,j}^{t-q}|) \bullet P(x_{i,j}^t|\boldsymbol{\Theta}_{a,b}). \qquad (1)$$

where $P(\boldsymbol{\Theta}_{a,b}|x_{i,j}^t, \cdots, x_{i,j}^{t-q})$ is the posterior probability of being classified as background and foreground based on $q+1$ surrounding frames, respectively. $P(\boldsymbol{\Theta}_{a,b}|x_{i,j}^{t-1}, \cdots, x_{i,j}^{t-q}|)$ is the prior distribution of background or foreground by incorporating information over q number of previous frames. $P(x_{i,j}^t|\boldsymbol{\Theta}_{a,b})$ is the likelihood distribution of the pixel with respect to the background or foreground model.

Based on Bayesian decision rule, a binary foreground detection map $M_{x,y}^t$ is

$$M_{i,j}^t = \begin{cases} 0, \text{ background, } P(\boldsymbol{\theta}_{a,b}^B|x_{i,j}^t, \cdots, x_{i,j}^{t-q}) \\ \qquad\qquad > P(\boldsymbol{\theta}_{a,b}^F|x_{i,j}^t, \cdots, x_{i,j}^{t-q}), \\ 1, \text{ foreground, } \text{ otherwise.} \end{cases} \qquad (2)$$

where $P(\boldsymbol{\theta}_{a,b}^B|x_{i,j}^t, \cdots, x_{i,j}^{t-q})$ and $P(\boldsymbol{\theta}_{a,b}^F|x_{i,j}^t, \cdots, x_{i,j}^{t-q})$ are the posterior probabilities of foreground and background, respectively.

The proposed Bayesian framework incorporates the spatial information $x_{i+k,j+m}^t$ of an individual pixel, $x_{i,j}^t$, the homogeneous information $R_{a,b}^c$ in a square block and temporal information $X_{a,b}$ to characterize the dynamic property of background and the statistical knowledge of a square block with time (as shown in Figure 2). Each homogeneous region in a square block is assumed nonstationary to describe disturbance at the background. The spatial information of an individual pixel captures the pixel's spatial movement within the block while the temporal information accumulates the statistical knowledge of homogeneous regions over time. The framework considers the dynamical property of background over short duration and the static property of background over long duration, therefore, the foreground false alarm is efficiently suppressed.

2.2   The Likelihood Probability

A likelihood probability reflects the degree-of-fit between pixel values and the corresponding background model and foreground model, respectively. In this paper, the dynamic property of background $X_{a,b}$ in a square block is modeled by a

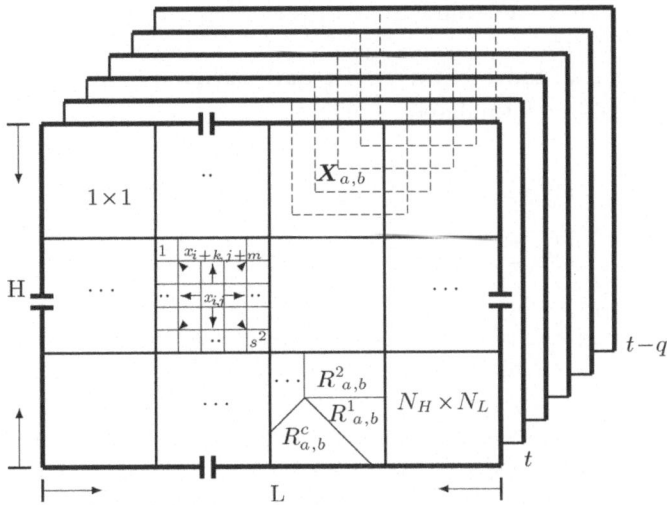

Fig. 2. Homogeneous region information, $R^c_{a,b}$ in a square block with size $s \times s$, spatial information, $x_{i+k,j+m}$ of pixels $x_{i,j}$, and temporal information $X_{a,b}$ from $t-q$ frame to t frame

set of homogeneous regions $R^c_{a,b}$. Based on the statistic analysis, it is reasonable to model each homogeneous region $R^i_{a,b}$, $i = 1, \cdots, c$, as single-Gaussian distribution, and a composition of different homogeneous regions forms a mixture of Guassian-like distribution in each $X_{a,b}$ denoted by $\boldsymbol{\theta}^B_{a,b} \sim \mathcal{N}(\boldsymbol{B}_{\mu_{R^i_{a,b}}}, \boldsymbol{B}_{\sigma_{R^i_{a,b}}})$ and $\boldsymbol{\theta}^F_{a,b} \sim \mathcal{N}(\boldsymbol{F}_{\mu_{R^i_{a,b}}}, \boldsymbol{F}_{\sigma_{R^i_{a,b}}})$, where $\boldsymbol{B}_{\mu_{R^i_{a,b}}}, \boldsymbol{B}_{\sigma_{R^i_{a,b}}}$, $\boldsymbol{F}_{\mu_{R^i_{a,b}}}$ and $\boldsymbol{F}_{\sigma_{R^i_{a,b}}}$ be the mean and the standard deviation of a homogeneous region $R^i_{a,b}$ of background scene and foreground object, respectively. Let $\mu_{R^i_{a,b}} = \{\mu^1_{R^i_{a,b}}, \cdots, \mu^d_{R^i_{a,b}}\}$ and $\sigma_{R^i_{a,b}} = \{\sigma^1_{R^i_{a,b}}, \cdots, \sigma^d_{R^i_{a,b}}\}$, where d is the dimension of feature space. In our implementation, we choose $d = 3$ for the color channels of a color camera.

The likelihood probability is obtained by the probability of each pixel belonging to a homogeneous region (foreground or background). In our work, the probability is directly proportional to the minimum distance between the pixel and every homogeneous region. It could be formulated as follows:

$$P(\boldsymbol{x}^t_{i,j}|\boldsymbol{\Theta}_{a,b}) = \{P(\boldsymbol{x}^t_{i,j}|\boldsymbol{\theta}^B_{a,b}), P(\boldsymbol{x}^t_{i,j}|\boldsymbol{\theta}^F_{a,b})\}, \tag{3}$$

$$P(\boldsymbol{x}^t_{i,j}|\boldsymbol{\theta}^B_{a,b}) = e^{-\lambda||D^{\boldsymbol{B}}{a,b}||^2}, \tag{4}$$

$$P(\boldsymbol{x}^t_{i,j}|\boldsymbol{\theta}^F_{a,b}) = \begin{cases} e^{-\lambda||D^F_{a,b}||^2}, & \text{if } e^{-\lambda||D^F_{a,b}||^2} \\ & > 1 - P(\boldsymbol{x}^t_{i,j}|\boldsymbol{\theta}^B_{a,b}), \\ 1 - P(\boldsymbol{x}^t_{i,j}|\boldsymbol{\theta}^B_{a,b}), & \text{otherwise.} \end{cases} \tag{5}$$

where $D^B_{a,b}$ and $D^F_{a,b}$ are minimum distances between $x^t_{i,j}$ and background model $\theta^B_{a,b}$ and foreground model $\theta^F_{a,b}$, respectively, λ is a factor to tune the degree-of-fit between the pixel value and the corresponding reference model to suit for different applications.

The minimum distance between $x_{x,y}^t$ and background or foreground model $\theta_{a,b}^B$ and $\theta_{a,b}^F$ could be formulated as follows:

$$
\begin{cases}
D_{a,b}^B = \arg\min_k \sum_{l=1}^d \left|(x_{i,j}^l - B_{\mu_{R_{a,b}^k}})\right|/B_{\sigma_{R_{a,b}^k}} \cdot \\
D_{a,b}^F = \arg\min_k \sum_{l=1}^d \left|(x_{i,j}^l - F_{\mu_{R_{a,b}^k}})\right|/F_{\sigma_{R_{a,b}^k}} \cdot
\end{cases}
\tag{6}
$$

By substituting equation (6) into equation (5), we can get the likelihood probability $P(x_{i,j}^t|\boldsymbol{\Theta}_{a,b})$ of the pixel $x_{x,y}^t$ to background model $\theta_{a,b}^B$ and foreground model $\theta_{a,b}^F$.

Note that in the case where there is no foreground model of a pixel at a particular time, the likelihood probability of the pixel belonging to the foreground model is calculated based on the probability of the pixel belonging to the background model. Thus, it is no need that the system must build foreground model at the initialization stage.

2.3   The Prior Probability

The prior distribution capturing similar information from the previous frames is a key component in facilitating accurate classification of pixels as either foreground or background. To represent dependencies between successive observations of a random variables especially if these observations are known to have some temporal signatures, Hidden Markov Model (HMM) is used. In this section, we will describe how a high order HMM with *mixture transition distribution* (MTD) could be used to capture information pertaining to similar features from the previous frames while preventing the parameter explosion of HMM.

In order to simplify the problem, we can rewrite equation (1) as follows:

$$
P(\boldsymbol{\Theta}_{a,b}|\boldsymbol{X}_{i,j}^n) \propto P(\boldsymbol{\Theta}_{a,b}|\boldsymbol{X}_{i,j}^{n-1}) \bullet P(x_{i,j}^t|\boldsymbol{\Theta}_{a,b}),
\tag{7}
$$

where

$$
P(\boldsymbol{\Theta}_{a,b}|\boldsymbol{X}_{i,j}^n) = P(\boldsymbol{\Theta}_{a,b}|x_{i,j}^t, \cdots, x_{i,j}^{t-q}),
\tag{8}
$$

$$
P(\boldsymbol{\Theta}_{a,b}|\boldsymbol{X}_{i,j}^{n-1}) = P(\boldsymbol{\Theta}_{a,b}|x_{i,j}^{t-1}, \cdots, x_{i,j}^{t-q}).
\tag{9}
$$

$P(\boldsymbol{\Theta}_{a,b}|\boldsymbol{X}_{i,j}^n)$ is the posterior distribution based on the observed points in $\{x_{i,j}^t, \cdots, x_{i,j}^{t-q}\}$ and $P(\boldsymbol{\Theta}_{a,b}|\boldsymbol{X}_{i,j}^{n-1})$ is the prior distribution based on observed points in $\{x_{i,j}^{t-1}, \cdots, x_{i,j}^{t-q}\}$.

In real-time application, since foreground and background feature distributions have a large overlap, it is not effective to detect foreground purely based on pixel feature discrepancy with the reference image. HMM facilitates a solution to this problem by first categorizing the pixels into "candidate" clusters, which are the *hidden variables*, $W = \{w_1, w_2\}$. The two mutually exclusive hidden states correspond to the different distinct parts of a monitored scene, such as background and foreground classes.

To compute the prior probability of a particular hidden state W sequence, equation (9) usually requires that we preserve all the points in $\boldsymbol{X}_{i,j}^{n-1}$ since the transition probability distribution among states is dependent upon the last q observations. Such qth-order Markov process is significantly more complex than the first order Markov hypothesis widely used. Unfortunately, as the order q of a Markov chain increases, the number of independent parameters increases exponentially and rapidly becomes too large to be estimated accurately with the data sets typically encountered in practice. The Mixture Transition Distribution (MTD) model [5] is introduced to approximate the high-order Markov chains with far fewer parameters than the fully parameterized model.

To construct the MTD model, an overlapping time-shifted window of size m (where m is smaller than q) is used to collect features from m frames into a buffer $F_{x,y}^{k}$.

$$\boldsymbol{F}_{i,j}^{k} = \{x_{i,j}^{t-1-k}, \cdots, x_{i,j}^{t-1-(k+m)}\}, k \in \mathbb{Z}. \tag{10}$$

The prior probability $P(\boldsymbol{\Theta}_{a,b}|\boldsymbol{X}_{i,j}^{n-1})$ can be calculated as

$$P(\boldsymbol{\Theta}_{a,b}|\boldsymbol{X}_{i,j}^{n-1}) = \sum_{k=1}^{q-m} P(k)P(\boldsymbol{\Theta}_{a,b}|\boldsymbol{F}_{i,j}^{k}), \tag{11}$$

where

$$P(\boldsymbol{\Theta}_{a,b}|\boldsymbol{F}_{i,j}^{k}) = \frac{1}{m}\sum_{k=1}^{m} P(\boldsymbol{\Theta}_{a,b}|x_{i,j}^{t-m}). \tag{12}$$

$P(k)$ is an exponential decay term giving more weight to more recent frames, i.e. $P(k) \propto e^{\lambda k}$ with λ having a negative value, the magnitude of which determines the degree of memory decay.

3   Updating Scheme

In real-time application, an on-line updating scheme is necessary to change the accumulated prior statistical knowledge whenever new evidence becomes available. We proposed on-line prior distribution learning and recursive updating the background model to adapt to environmental changes.

3.1   On-line Prior Distribution Learning

The prior distribution of the surrounding frames is statistical properties of the given pattern source over a long period. An on-line prior distribution learning process establishes a recursive estimation of the prior probability by taking into account the new incoming observation $x_{i,j}^{t}$ and its posterior distribution in the previous frame. In our paper, the main idea of the on-line prior distribution learning is to update the prior probability of the current frame by minimizing the mean square error between the estimated prior knowledge of next frame and

the posteriy distribution of the current frame over a short frame buffer. The
probability updating guarantees the accuracy of the accumulated prior probabilities and continuous detection.

Suppose

$$y = \{P(\boldsymbol{\Theta}_{a,b}|\boldsymbol{X}_{i,j}^n), \cdots, P(\boldsymbol{\Theta}_{a,b}|\boldsymbol{X}_{i,j}^{n-k}), \cdots, P(\boldsymbol{\Theta}_{a,b}|\boldsymbol{X}_{i,j}^{n-m})\} \tag{13}$$

be the posterior distribution based on an observed point set from $\boldsymbol{X}_{i,j}^n$ to $\boldsymbol{X}_{i,j}^{n-m}$,
and

$$x = \{P(\boldsymbol{\Theta}_{a,b}|\boldsymbol{X}_{i,j}^{n-1}), \cdots, P(\boldsymbol{\Theta}_{a,b}|\boldsymbol{X}_{i,j}^{n-k-1}), \cdots, P(\boldsymbol{\Theta}_{a,b}|\boldsymbol{X}_{i,j}^{n-m-1})\}. \tag{14}$$

be the prior distribution based on an observed point set from $\boldsymbol{X}_{i,j}^{n-1}$ to $\boldsymbol{X}_{i,j}^{n-m-1}$.
Both are random variables in $\mathcal{N}(\boldsymbol{\mu_y}, \boldsymbol{\sigma_y})$ and $\mathcal{N}(\boldsymbol{\mu_x}, \boldsymbol{\sigma_x})$, respectively. The task
is then to find an estimator $\hat{x} = \Phi(x)$, which is an affine function $\hat{x} = Kx + c$,
as a recursive scheme, such that

$$\hat{x} = \Phi(x) \text{ is close to } y. \tag{15}$$

This can be achieved by choosing Φ which minimizes the mean square error
$E\|\Phi(x) - y\|$ to update the prior distribution.

The estimation error is here:

$$z = \Phi(x) - y \tag{16}$$

$$= Kx + c - y \tag{17}$$

$$= [K \; -I] \begin{bmatrix} x \\ y \end{bmatrix} + c, \tag{18}$$

so we have

$$cov(z) = [K \; -I] \begin{bmatrix} \boldsymbol{\sigma}_x & \boldsymbol{\sigma}_{xy} \\ \boldsymbol{\sigma}_{yx} & \boldsymbol{\sigma}_y \end{bmatrix} \begin{bmatrix} K^T \\ -I \end{bmatrix}. \tag{19}$$

We would like to minimize

$$E\|z\|^2 = trace(cov(z)). \tag{20}$$

Thus, the mean square error is

$$E\|z\|^2 = trace\left(\sigma_y - \sigma_{yx}\sigma_x^{-1}\sigma_{xy}\right)$$
$$+ trace\left((\boldsymbol{K} - \sigma_{yx}\sigma_x^{-1})\sigma_x(\boldsymbol{K} - \sigma_{yx}\sigma_x^{-1})^T\right). \tag{21}$$

Since the matrix, $trace\left((\boldsymbol{K} - \sigma_{yx}\sigma_x^{-1})\sigma_x(\boldsymbol{K} - \sigma_{yx}\sigma_x^{-1})^T\right)$, is positive semidef-
inite, so its trace,which is the sum of the eigenvalue, is always positive.

Thus, the $\boldsymbol{K}$ which minimizes the mean square error is

$$\boldsymbol{K} = \sigma_{yx}\sigma_x^{-1}. \tag{22}$$

$$c = \mu_y - \boldsymbol{K}\mu_x. \tag{23}$$

The prior knowledge is updated as follows:

$$\hat{x} = \mu_y + \sigma_{yx}\sigma_x^{-1}(x - \mu_x). \tag{24}$$

Finally, the prior probability of $\boldsymbol{x}_{i,j}^t$ is updated,

$$\hat{P}(\boldsymbol{\Theta}_{a,b}|\boldsymbol{x}_{i,j}^t) \Longleftarrow \mu_y + \sigma_{yx}\sigma_x^{-1}(x - \mu_x). \tag{25}$$

3.2   Updating Background Model

After collecting a number of pixels belonging to the background model $\theta_{a,b}^{B}$, parameters $\boldsymbol{B}_{\mu_{R_{a,b}^{k}}}$ and $\boldsymbol{B}_{\sigma_{R_{a,b}^{k}}}$ at time-t are updated as follows:

$$\boldsymbol{B}_{\mu_{R_{a,b}^{k}}^{t}} \Longleftarrow (1 - \rho)\boldsymbol{B}_{\mu_{R_{a,b}^{k}}^{t-1}} + \rho\boldsymbol{B}_{\mu_{R_{a,b}^{k}}^{t}}, \tag{26}$$

$$\boldsymbol{B}_{\sigma_{R_{a,b}^{k}}^{t}} \Longleftarrow (1 - \rho)\boldsymbol{B}_{\sigma_{R_{a,b}^{k}}^{t-1}} + \rho\boldsymbol{B}_{\sigma_{R_{a,b}^{k}}^{t}}. \tag{27}$$

where $\rho = 1/T$ is the learning factor for adapting the current environmental change, and T is the number of "clean" background frames collected for constructing the initial background model. The above shows a recursive linear interpolation scheme for continuously updating background models, which shares similarity as the scheme used in [6].

4   Experimental Analysis

Extensive experiments have been carried out to evaluate the performance of the proposed algorithm over long operational hours under a variety of environments, e.g, the office scene with moving curtains,the shopping mall, the campus scene with moving tree branches and the swimming pool at nighttime with point-light reflections, as shown in Figure 3.

The difficulties for foreground detection in the office can be caused by moving curtains and partly hidden foreground objects (i.e., the color of the foreground object is similar to that of the covered background) shown in the 1st row of Figure 3. In the campus scene, changes of the background are often caused by the motion of tree branches and their shadows on the ground surface shown in the 5th row of Figure 3. Due to continual disturbances caused by water ripples and splashes, in the swimming pool, the dynamic property of the background is relatively more hostile than most indoor and outdoor environments. The background movements as the reflective regions and lane dividers could be easily misidentified as foreground objects' movements as shown in the 7th row of Figure 3. In addition, poor visibility of swimmers in water due to reflections (from sunlight and nighttime lighting) results in highly fragmented or even missing foreground. As shown in the even rows of Figure 3, foreground objects are robustly detected even when the objects go through regions hampered by reflections.

Figure 4 shows the comparison results of our proposed algorithm with spatial-temporal-based approach [2] and traditional frame-based method, such as the pixel-based Mixture Gaussian Model method (MGM) [6]. It can be seen that our proposed algorithm consistently achieves better object detection results in terms of the capability to detect small objects and suppressing the errors due to dynamic background.

To quantitatively compare the performance of MGM and our proposed algorithm, we adopted the 'detection failure rate' (DFR) measure [7] as follows:

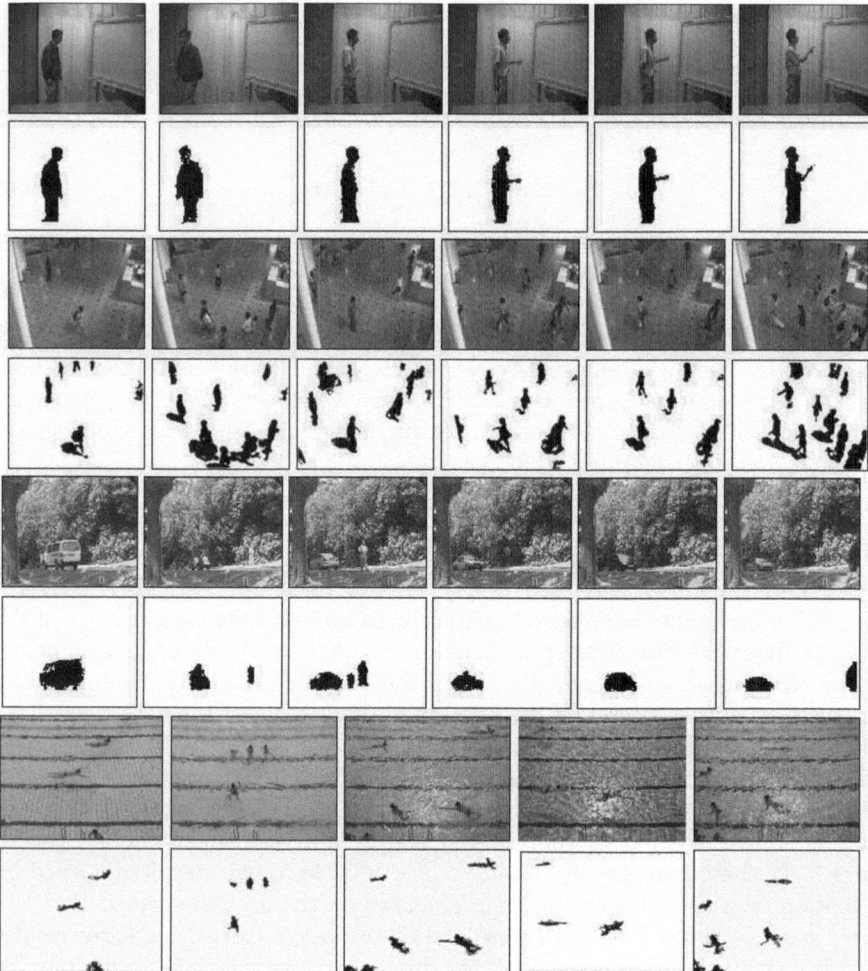

Fig. 3. Experimental results under different environments, i.e., the office with moving curtain, the shopping mall, the campus with moving tree and the swimming pool with dynamic background and specular reflectance

$$\text{DFR} = \frac{\text{Number of detection failure foreground}}{\text{Number of benchmark of foreground}} \qquad (28)$$

Table 1 shows the comparison results obtained within hostile aquatic environments. For instance, normal swimming video, daytime swimming video with reflection, swimming video with reflection at nighttime, treading video and distress video. We can see that the foreground detection rate of the proposed algorithm is better than pixel-based MGM algorithm under noisy aquatic environments. Especially, our proposed algorithm has an advantage detecting long-term static foreground where the swimmer is treading and under distress.

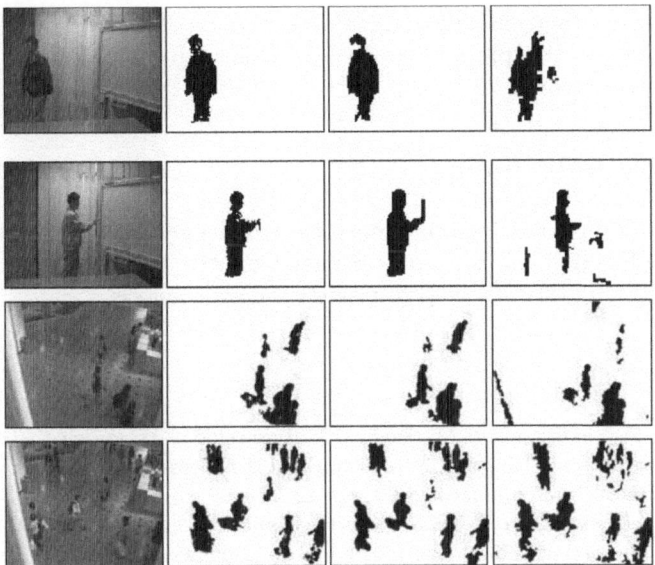

Fig. 4. Comparison results of the proposed algorithm with spatial-temporal-based approach [2] and the pixel-based Mixture Gaussian Model method (MGM) [6], respectively. 1st column: Samples of scene captured; 2nd column: Segmented objects using our proposed algorithm; 3rd column: Segmented object using spatial-temporal based approach and 4th column: Segmented object using MOG

Table 1. Quantitative evaluation of foreground detection in term of detection failure rate (DFR)

No.	Video	Pixel-based Mixture Gaussian Model	Proposed method
		DFR	DFR
1	Normal swimming video	6%	5%
2	Daytime with reflection	43%	6%
3	Nighttime with reflection	46%	10%
4	Treading video	45%	2%
5	Distress video	35%	6%

5   Conclusion

In this work, a framework for foreground subtraction applicable for various complex environments is proposed. This framework jointly incorporated the spatial, spectral and temporal information of the video sequence. This framework improved the foreground subtraction results by utilizing the accumulated statistical information over time based on the Bayesian approach. In addition, a novel recursive on-line prior distribution learning process is proposed to minimize the difference between the prior distribution and posterior distribution of pixels, while guaranteeing the accuracy of the prior distribution accumulation.

Compared with other existing change detection methods under different complex environments, this proposed approach achieved better detection results and lowest detection error rate.

6   Acknowledgements

This work is supported by the Enterprise Challenge Unit, Singapore's Prime Minister's Office and the Singapore Sports Council. The authors would like to thank Dr Li Liyuan for his help to provide video sequences.

References

1. O. Javed, K. Shafique and M. Shah, "A Hierarchical Apporach to Robust Background Subtraction Using Color and Gradient", *Proceddings of IEEE Workshop on Motion and Video Computing*, 2002, pp. 22–27.
2. L. Li, W. M. Huang, I. Y. H. Gu and Q. Tian, "Foreground Object Detection in Changing Background based on Color Co-occuranence Statistics", *Proceddings of IEEE Workshop on Applications of Computer Vision*, 2002, pp. 269–274.
3. G. L. Foresti. "Face Detection and Tracking in a Video by Propagating Detection Probabilities", *IEEE Trans. Pattern Anal. and Machine Intell.*, 2003, no. 10, vol. 25, pp. 1215– 1228.
4. H.-L. Eng, K.-A. Toh, A.H. Kam, J. X. Wang and W.-Y. Yau, "An Automatic Drowning Detection Surveillance System for Challenging Outdoor Pool Environments" *Proceddings of IEEE International Conference on Computer Vision*, 2003, pp. 532–539.
5. A. Raftery and S. Tavare. "Estimation and Modelling Repeated Patterns in High Order Markov Chains with the Mixture Transition Distribution Model", *Applied Statistics*, 1994, no. 43, vol. 1, pp. 179–199.
6. C. Stauffer and L.Grimson, "Learning Patterns of Activity Using Real-time Tracking", *IEEE Trans. Pattern Anal. Machine Intell.*, 2000, no. 22, vol. 8, pp. 747–757.
7. S. C. Liu, C. W. Fu and S. Chang. "Statistical Change Detection with Moments Under Time-varying illumination", *IEEE Trans. Image Processing*, 1998, no. 7, vol. 9, pp. 1258–1268.

A Background Maintenance Model in the Spatial-Range Domain

Daniel Kottow, Mario Köppen, and Javier Ruiz-del-Solar

Fraunhofer IPK, Dept. Pattern Recognition, Pascalstr. 8-9, 10587 Berlin, Germany
DIE, U. de Chile, Tupper 2007, Santiago, Chile
dk@poroto.net, {mario.keoppen.ipk.fhg.de, jruizd}@cec.uchile.cl

Abstract. In this article a background maintenance model defined by a finite set of codebook vectors in the spatial-range domain is proposed. The model represents its current state by a foreground and a background set of codebook vectors. Algorithms that dynamically update these sets by adding and removing codebook vectors are described. This approach is fundamentally different from algorithms that maintain a background representation at the pixel level and continously update their parameters. The performance of the model is demonstrated and compared to other background maintenance models using a suitable benchmark of video sequences.

1 Introduction

Many video surveillance applications (e.g. people tracking) rely on a background maintenance model which is able to segment moving foreground objects from an essentially stationary background image. These models must deal with a variety of distracting intensity changes in the image. Toyama et al. have made a detailed description of these intensity variations, classified them in a set of canonical problems and introduced benchmark video sequences for them in [1].

Most existing algorithms model the intensity variations in the background by statistical models at the pixel level; the mixture of gaussian models presented in [6] is a prominent example thereof. We model the intensity values belonging to the background by a finite set of codebook vectors in the so-called spatial-range domain, whic was introduced by Comaniciu et al. in [2]. Intensity values in the image are either close enough to a background codebook vector to be considered background or are classified as foreground. By using the spatial-range domain time-varying background intensities at one pixel location can be modeled by several vectors sharing the same spatial coordinate; on the other hand, a single codebook vector can represent several pixels if they have roughly the same intensity value.

We present the spatial-range domain and define an algorithm for representing an image frame by a set of codebook vectors in section 2. In section 3 we define algorithms which update a background set of codebook vectors in order to maintain an up-to-date background representation. In section 4 we demonstrate the

D. Comaniciu et al. (Eds.): SMVP 2004, LNCS 3247, pp. 141–152, 2004.

performance of our model on the benchmark problems presented in [1]. Finally, in section 5 some conclusions and projections of this work are given.

2 Vector Quantization of an Image in the Spatial-Range Domain

The ability to process images taking into account simultaneously the location and the intensity value of a pixel has been successfully exploited by an edge preserving filtering technique known as bilateral filtering [7] and also by the mean shift algorithm for image segmentation [2] presented by Comaniciu et al.

As Comaniciu et al. put it, "*an image typically is represented as a 2-dimensional lattice of r-dimensional vectors, where e.g. r = 1 for the gray level case and r = 3 for the color case. The space of the lattice is known as the spatial domain of the image, while the pixel values are in the range domain. However, after a proper normalization, the location and range vectors can be concatenated to obtain a spatial-range domain of dimension d = r + 2.*" [2].

We adopt this approach and represent an image as a set of vectors in the spatial-range domain:

$$I = \{x_j\}_{j=1..n} \tag{1}$$

Each vector x_j has two parts, a spatial and a range part, where the range part may be written as a function of the spatial part:

$$x_j = (x_j^s, x_j^r) = (x_j^s, I(x_j^s)) \tag{2}$$

The superscripts s and r denote the spatial and range parts of the vectors, respectively. We will refer to the image vectors as pixels, and name the spatial and the range part, pixel location and pixel value, respectively.

We want to encode an image by a set of codebook vectors $C = \{c_i\}_{i=1..m}$ in the spatial-range domain. Each codebook vector will represent an equally-sized portion of the spatial-range domain, namely, all vectors lying inside a constant-sized (hyper)rectangle $\bar{c}_i$ centered at the codebook vector:

$$x \in \bar{c}_i \Leftrightarrow \|x^s - c_i^s\| < \sigma_s \wedge \|x^r - c_i^r\| < \sigma_r \tag{3}$$

The dimension of the rectangle is given by σ_s and σ_r, which are constant and independent of i. They define the accuracy of the representation in each domain and by that, introduce normalization factors which compensate for the different magnitudes found in spatial and range data.

The representation of an image $I = \{x_j\}_{j=1..n}$, requires a set of codebook vectors $C = \{c_i\}_{i=1..m}$ which cover all the image pixels:

$$\forall x_j \in I \, \exists c_i \in C \, (x_j \in \bar{c}_i) \tag{4}$$

A trivial representation of an image is given by the image pixels themselves. However, this is a highly redundant set in the sense of eq. 4; to make the representation more compact we can suitably subsample the image. Algorithms 1 and 2 follow this strategy. Figure 1 shows an image of Lenna, a set of codebook vectors obtained by algorithm 2 and an image illustrating the representation quality of the codebook vectors. To obtain this image, each pixel value was replaced by the color value of the codebook vector that represents it. We achieve a coding compression, given by the ratio between the number of pixels and the number of codebook vectors, of aproximately 1:15. This saves computation time when we locally process images at the codebook vector level instead of the pixel level, as we do in the background model presented in section 3.

Algorithm 1 $Learn(x, C)$ //make C represent image vector x

if $!\exists c_i \in C\,(x \in \bar{c}_i)$:
   $C \leftarrow C \cup \{x\}$

Algorithm 2 $LearnImage(I, C)$ //make C represent image I

foreach $x_j \in I$:
   $Learn(x_j, C)$

Fig. 1. The left image shows a picture of Lenna. In the middle, a set of codebook vectors to represent it. The right image illustrates the representation quality of the codebook vectors by replacing each pixel value by the greyscale value of the codebook vector that represents it

3   Background Model

Our background model represents the background by a set of codebook vectors in the spatial-range domain; image pixels not represented by this set will be output as the segmented foreground. To cope with the intensity variations of the background mentioned in the introduction this set of codevectors is continously updated. In this section we present the background maintenance model consisting of the set of background codebook vectors, another set of codebook vectors representing the foreground, an algorithm that decides at the frame-level whether an object has been detected, and several mechanisms that update the background and foreground set of codebook vectors by adding and removing image pixels in order to obtain an accurate foreground segmentation.

3.1   Foreground Background Representation

The model needs to keep information about the current and recent past foreground. There are mainly two reasons for this. First, because static objects in the foreground should become background after a reasonable time, and, second, to take frame-level based decisions by integrating properties of the foreground codevectors as will be described in section 3.3.

We use two sets of codebook vectors, one for the background information and the other for the foreground. Given an image pixel x and the codebooks $C_{bg} = \{c_{i1}\}$ (background) and $C_{fg} = \{c_{i2}\}$ (foreground) we define $DualLearn$ (alg. 3), which tells us if and which set of codebook vectors represents the image pixel, and if none does, it adds the pixel to the foreground.

Algorithm 3 $DualLearn(x, C_{bg}, C_{fg})$ //make C_{fg} represent x if C_{bg} does not

---

```
if ∃c_{i1} ∈ C_{bg} (x ∈ c̄_{i1}):
    return c_{i1}
else if ∃c_{i2} ∈ C_{fg} (x ∈ c̄_{i2}):
    return c_{i2}
else:
    C_{fg} ← C_{fg} ∪ {x}
    return x
```

With the purpose of managing computational time effectively, we define a seed-growing strategy based on $DualLearn$. Given a set of image pixels $S = \{x_i\}$, $DualGrow$ (alg. 4) detects and processes areas where novel image content is found, while skipping areas that remain unchanged.

Algorithm 4 $DualGrow(I, S, C_{bg}, C_{fg})$ //make C_{fg}, C_{bg} represent S and connected regions

---

```
while S ≠ ∅:
    x ← PopItem(S)
    y ← DualLearn(x, C_{bg}, C_{fg})
    if y = x:  //x was added to foreground
        foreach x_n^s ∈ ConnectedNeighbors(x^s):
            PushItem(S, (x_n^s, I(x_n^s)))
```

$PopItem(S)$ removes and returns the first element from the ordered set S. $PushItem(S)$ adds an element to the end of S. $ConnectedNeighbors(x^s)$ returns the 4-connected neighbors of x^s in the two-dimensional lattice space of the image.

3.2 Dynamical Lifetime for Codebook Vectors

Maintaining an accurate background set of codebook vectors for an image sequence $\{I_t\}_{t=0..T}$ requires a method for removing codevectors when changes in the background occur. We introduce a dynamical attribute s_i called score for every codevector c_i. If the codevector represents the current image pixel at the codevectors spatial location, the score increases, otherwise it decreases:

$$s_i(t+1) = \begin{cases} (1-\tau)s_i(t) + \tau & if \ \|c_i^r - I_t(c_i^s)\| < \sigma_r \\ (1-\tau)s_i(t) - \tau & otherwise \end{cases} \tag{5}$$

This equation corresponds to a simple diffusion process using the outcome of the present representation ability as an exogenous input. s_i is initialized with 0; τ controls the rate of change of s_i. Codevectors with a score below some threshold are considered obsolete and removed. On the other hand, a maximum score is used to identify static foreground. Details are given in section 3.5.

3.3 Frame-Level Based Decisions

Our background maintenance model integrates information from the foreground set of codevectors that is used to discriminate between object detections and global illumination changes. We classify the foreground set states into four situations:

1. The set of foreground codevectors is small
2. The set of foreground codevectors covers most parts of the image
3. The set of foreground codevectors is spatially sparse
4. Any other case

If any of the first three situations happened, the detection is considered a false alarm (for different reasons) and the background is adapted to the changes in illumination captured by the foreground set as will be detailed in section 3.5. Otherwise, an object detection event is triggered. We now describe this situations in more detail, the formal representations for them are given in the algorithm *Detection* (alg. 5).

1. A small set of foreground codevectors may be due to noise, small, unstable regions of the image or other local phenomena. The actual decision boundary, N_{noise}, is chosen by the user and should be in balance with the size of the objects to be detected.

   But also slowly varying, global, lighting conditions produce higher noise levels and therefore comply with this case. When a light change takes place, it does not change the intensity levels of all image pixels by the same magnitude. Furthermore, due to the tolerance in the intensity representation as given by eq. 3, at all times each codevector matches the current video frame with a particular, local precision. If the scene undergoes a steady and smooth light change, mismatches in the background representation will be more frequent but will extend over the entire period of time of the light change and therefore, at each frame, only a few codevectors spread over the image are produced as foreground.

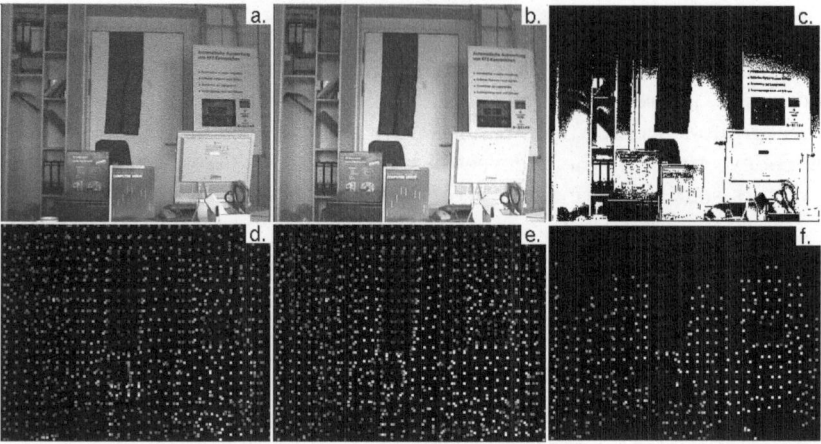

Fig. 2. Images (a) and (b) show our lab under different lighting conditions and are represented by the codebook vectors shoewn in (d) and (e), respectively. In both cases the codebook vector density in the spatial domain is $1.8\sigma^s$. Image (c) shows a thresholded difference image between both scenes illustrating that edges are more robust to light changes than reflecting surfaces. We applied the algorithm *DualGrow* to the lighter scene (b) using the codebook vectors (d) representing the darker scene as background. The resulting foreground codebook vectors are shon in image (f). Their density is $1.3\sigma^s$

2. The set of foreground codevectors is spread over (almost) all the spatial range of the image, we assume this to happen because of a global, sudden illumination change or an object that crossed right in front of the camera.
3. The third condition is intimately related to the usage of the spatial-range domain. Projecting the foreground vectors onto the image spatial plane, and leaving out areas where no codevectors occur, we calculate the mean density of vectors. The foreground codebook vectors are considered spatially sparse if the mean density in the spatial domain is below some user-defined value. If the foreground vectors are spatially sparse, the system will not output an object detection. To understand the underlying motivation for this rule, consider that coding of uniformally colored surfaces in the spatial-range domain is less dense than coding highly structured image portions such as edges and textures, as is readily observed in figure 1. We have performed some experiments suggesting that an illumination change that is not strong enough to affect the whole image, is mostly reflected in intensity variations for uniformally colored, reflecting surfaces, while object edges and texturized surfaces are less affected. An example is shown in figure 2. Other work [4], which takes into account texture and edge differences between a background reference and the image rely on the same hypothesis. We plan to study the relation between an illumination change and the affected image areas in more detail in future work.

Algorithm 5 $Detection(C_{fg})$ //classify detected foreground

```
if |C_fg| < N_noise:
    return NOISE
```
$$\text{else if } \frac{I^s - \bigcup_i D(c_i^s)}{I^s} > A_{global}:$$
```
    return GLOBAL
```
$$\text{else if } \frac{\sum_i N(c_i^s)}{|C_{fg}|} < \delta_{sparse}:$$
```
    return SPARSE
else:
    return OBJECT
```

I^s denotes the spatial extent of the image (image size). $D(c_i^s)$ returns the area of a disc of radius σ_s centered at c_i^s. $N(c_i^s)$ returns the number of cells contained in $D(c_i^s)$. $\bigcup_i$ and $\sum_i$ iterate over the foreground codevectors C_{fg}.

3.4   A Perceptually Uniform Color Space and a Low-Pass Filter

The range part of the image pixels in our system comes from color images. We have been using mainly two color spaces to represent color values, RGB and the perceptually uniform CIE color space Lab. The CIE characterizes color using one luminance parameter (L) and two chromaticity variables (ab). Consequently, we define two normalization factors within the range domain, namely, a luminance normalization factor σ_{rl} and a chromaticity normalization factor σ_{rc}. This modifies eq. 3 to read:

$$x \in \bar{c}_i \Leftrightarrow \left\| x^s - c_i^s \right\| < \sigma_s \wedge \left\| x^{rl} - c_i^{rl} \right\| < \sigma_{rl} \wedge \left\| x^{rc} - c_i^{rc} \right\| < \sigma_{rc} \qquad (6)$$

In order to smooth the foreground detection, we also low-pass filter the images with a small averaging filter before processing them. Another way of looking at the filtering as being part of the vector coding process of the images is given in [3].

3.5   Tracey: Putting It All Together

Now that we have defined all the modules involved in our background model, we integrate them to a complete background maintenance model called *Tracey*. The first few frames are processed with *LearnImage* (alg. 2), which initializes the system. Subsequent frames are processed with *RecallImage* (alg. 6). We now give a verbal description of *RecallImage*, which is the core of our background maintenance model:

1. We sample some image pixels covering all the spatial image extent in a uniform manner. These pixels are fed into *DualGrow* (alg. 4) which adds new foreground codevectors where novel image content is detected.
2. The foreground classifier *Detection* (alg. 5) determines whether the foreground vectors represent an object or some task-irrelevant intensity changes.

3. If an object is detected, the set of foreground vectors is maintained. *UpdateForeground* (alg. 8) checks which foreground vectors have a low score and therefore are supposed to represent a static object (s_i larger than s_{static} threshold), in which case they become background. On the other hand, if no object is detected, the foreground set of codevectors induces background learning and is then erased.

4. *UpdateBackground* (alg. 7) checks the validity of the background set and removes obsolete codevectors (s_i smaller than s_{death} threshold).

The dynamic nature of our model is ultimately given by the addition and removal of codebook vectors. This discrete, non-linear way of adaptation to all kinds of time-varying phenomena in the background is in sharp contrast to models that update a representation at the pixel level by continously changing real-valued parameters as in [6, 1].

We have run the background maintenance model with an image resolution of 320x256 pixels and obtained a frame rate of 15-20 fps on a PC with a 1Ghz processor.

Algorithm 6 $RecallImage(I, C_{bg}, C_{fg})$ //process image frame

---
$S \leftarrow SampleSpace(I)$
$DualGrow(I, S, C_{bg}, C_{fg})$
$d \leftarrow Detection(C_{fg})$
if $d = OBJECT$:
    $UpdateForeground(I, C_{bg}, C_{fg})$
else:
    foreach $c_i^s \in C_{fg}$:
        $Learn(c_i^s, C_{bg})$
    $C_{fg} \leftarrow \emptyset$
$UpdateBackground(I, C_{bg})$

$SampleSpace(I)$ returns a set of image pixels from I which cover the complete spatial image extent.

Algorithm 7 $UpdateBackground(I, C_{bg})$ //check for removal of background

---
foreach $c_i \in C_{bg}$:
    $x \leftarrow (c_i^s, I(c_i^s))$
    if $x \in \bar{c}_i$:
        $s_i \leftarrow (1 - \tau)s_i + \tau$
    else:
        $s_i \leftarrow (1 - \tau)s_i - \tau$
        if $s_i < s_{death}$:
            $C_{bg} \leftarrow C_{bg} \setminus \{c_i\}$

Algorithm 8 $UpdateForeground(I, C_{bg}, C_{fg})$ //check for static objects in foreground

> foreach $c_i \in C_{fg}$:
>> $x \leftarrow (c_i^s, I(c_i^s))$
>> if $x \in \bar{c}_i$:
>>> $s_i \leftarrow (1 - \tau)s_i + \tau$
>>> if $s_i > s_{static}$:
>>>> $C_{bg} \leftarrow C_{bg} \cup \{x\}$
>>>> $C_{fg} \leftarrow C_{fg} \setminus \{c_i\}$
>> else:
>>> $C_{fg} \leftarrow C_{fg} \setminus \{c_i\}$

4   Simulation Results

In order to assess the performance of our algorithms, we decided to run the so-called canonical background problems presented by Toyama et al. in [1]. These video sequences were taken from different settings, each one presenting another kind of difficulty a background model should cope with. We have run the benchmarks with three different flavors of our background model.

1. *Tracey Lab LP* treats all image data in the CIE Lab color space and low-pass filters the frames with a 2x2 mean filter before further processing.
2. *Tracey RGB LP* treats all image data in the RGB color space and low-pass filters the frames with a 2x2 mean filter before further processing.
3. *Tracey RGB* treats all image data in the RGB color space. There is no preprocessing.

We run our algorithm in these three different variants, because that shows which aspects of the performance are due to the algorithm itself, and what may be gained by a different color space or smoothing of the image. For comparison reasons, we reproduce the results obtained and published in [1] for the three, in our opinion, most successful approaches described there. The foreground segmentation obtained at the evaluation frame are shown in figure 3 and the corresponding classification errors are summarized in table 1.

We now give a brief description of each video sequence up to the frame where the foreground segmentation is evaluated. A more detailed description can be found in the original publication.

- Moved Object (MO): A person comes into a room, sits down and goes out again. The chair is left in a new position.
- Time of Day (ToD): The lights gradually get brighter. Then a person walks into the room and sits down.
- Light Switch (LS): A room with the lights on, then they are switched off for a long time. A person enters the room, switches on the light and sits down while the door closes.
- Waving Trees (WT): A tree is waving in the wind and a person steps in front of it.

Table 1. Numerical results of foreground segmentation on Wallflower benchmark. We reproduce the results published in [1] for comparison reasons. TE stands for total errors and is the sum of the false positives and false negatives across all video sequences. The acronyms for the benchmarks are explained in the text

		MO	ToD	LS	WT	C	B	FA	TE	TE*
Tracey Lab LP	f. pos.	1	54	2024	136	69	92	356		
	f. neg.	0	772	1965	191	1998	1974	2403	12035	8046
Tracey RGB LP	f. pos.	0	19	7521	152	1058	262	602		
	f. neg.	0	1071	2196	252	875	1784	2627	18419	8702
Tracey RGB	f. pos.	0	20	7509	131	1025	248	663		
	f. neg.	0	1133	2344	390	1329	1804	2672	19268	9415
Wallflower	f. pos.	0	961	947	877	229	2025	320		
	f. neg.	0	25	375	1999	2706	365	649	11478	10156
Eigenbackground	f. pos.	1065	16	362	2057	1548	6129	537		
	f. neg.	0	879	962	1027	350	304	2441	17677	16353
Mixture of Gaussians	f.pos.	0	20	14169	341	3098	217	530		
	f. neg.	0	1008	1633	1323	398	1874	2442	27053	11251

*without light-switch

- Camouflage (C): A person comes in and stands in front of a flickering monitor which is displaying, among other colors, a very similar color to the person's shirt.
- Bootstrapping (B): A busy cafeteria with people constantly showing up, the scene is never devoid of people.
- Foreground Aperture (FA): A person is shown from behind and starts moving to the right. Due to lighting the person appears entirely black.

Our algorithm succeeds on the most common problems in background modeling: incorporating static objects in the background (Moved Object), handling time-varying textures (Waving Trees, Camouflage), adapting to gradual light changes (Time of Day) and being able to function without a clean background reference (Bootstrapping). It particularly excels where the intensity variations of the background follow a complex time pattern, like in Waving Trees. On the other hand, since our background model does not explicitly represent the time-variations for the pixel values itself, it cannot detect the suddenly constant intensity that occurs in the Camouflage benchmark. Although our system is capable of handling several sudden light changes in our lab, the algorithm performs rather poorly on the particular benchmark Light Switch. The successful algorithms on this benchmark had learned the lighting condition given at the beginning of the video and remembered it when the light was switched on again. Our algorithm has no long-term memory to do so. Finally, a problem like Foreground Aperture requires a region-level processing algorithm, something that is particular to the Wallflower model and is seen as a post-processing step by the authors themselves. Despite the fact that our model does not explicitly model these phenomena, the results obtained by all three variants of *Tracey* are highly

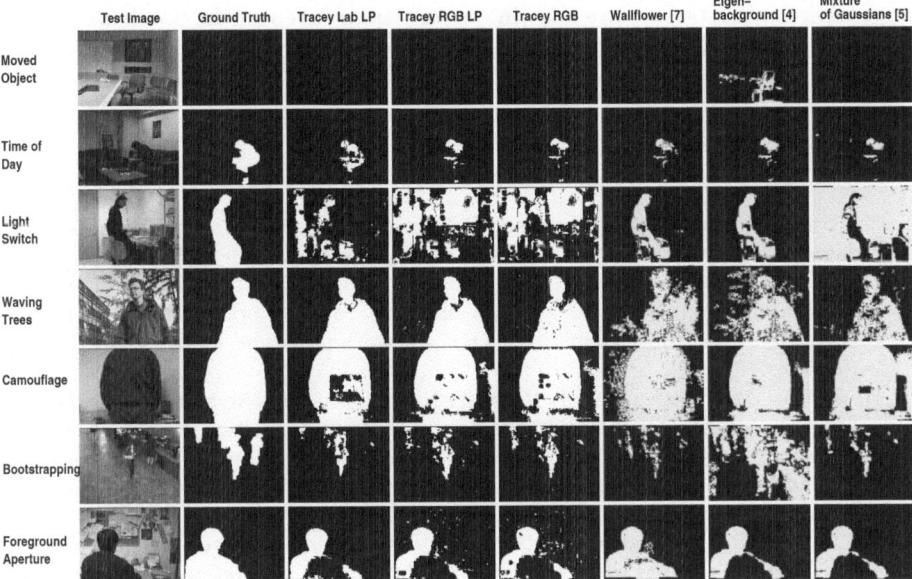

Fig. 3. Foreground segmentation for the Wallflower benchmark. Each row shows the results of each algorithms when applied to one of the seven canonical problems defined in [1] and are publicly available on the Internet. The test image corresponds to the frame in the video sequence when the evaluation of the segmentation takes place. The Ground Truth is a hand segmented reference. *Tracey Lab LP* , *Tracey RGB LP* and *Tracey RGB* are three variants of our background maintenance model as described in the text. Wallflower [1], Eigenbackground [5] and Mixture of Gaussians [6] are described in the corresponding reference. We reproduce the results published in [1] for comparison reasons

competitive; in fact, when excluding the Light-Switch scenario they outperform all other algorithms tested by Toyama et al (see table 1).

5 Conclusions

We presented an algorithm for background maintenance based on a set of codebook vectors in the spatial-range domain. While most algorithms work at the pixel level by proposing a statistical model for each pixel value, our approach uses simple dynamical rules to add or remove prototypes that represent local background intensities. The model can in principle represent arbitrary complex time-varying image intensities. A frame-level analysis discriminates between global light changes, noise and objects of interest and updates the background representation accordingly. We evaluated our background maintenance model on a suite of canonical benchmarks and obtained results comparable to the best approaches published before showing a particular strength for modeling quasi-periodical local light changes as given by speculating water surfaces or waving trees.

References

1. K. Toyama J. Krumm B. Brumitt and B. Meyers. Wallflower: Principles and practice of background maintenance. In *Proceedings of the IEEE International Conference on Computer Vision*, 1999.
2. D. Comaniciu and P. Meer. Mean shift: A robust approach toward feature space analysis. *IEEE Transactions on Pattern Analysis and Machine Intelligence*, 24(5):603–619, 2002.
3. D. Kottow M. Köppen and J. Ruiz del Solar. Temporal dynamical interactions between multiple layers of local image features for event detection in video sequences. In *Proceedings of the 13th Scandinavian Conference on Image Analysis*, 2003.
4. L. Li and M. Leung. Integrating intensity and texture differences for robust change detection. *IEEE Transactions on Image Processing*, 11(2), 2002.
5. N. Oliver B. Rosario and A. Pentland. A bayesian computer vision system for modeling interactions. In *Proceedings of IEEE International Conference on Computer Vision*, 1999.
6. C. Stauffer and W.E.L. Grimson. Learning patterns of activity using real-time tracking. *IEEE Transactions on Pattern Analysis and Machine Intelligence*, 22(8):747–757, 2000.
7. C. Tomasi and R. Manduchi. Bilateral filtering for gray and color images. In *Proceedings of the IEEE International Conference on Computer Vision*, 1998.

A New Robust Technique for Stabilizing Brightness Fluctuations in Image Sequences

François Pitié, Rozenn Dahyot, Francis Kelly, and Anil Kokaram

Electronic and Electrical Engineering Department,
Trinity College Dublin, Ireland
{fpitie, dahyotr, frkelly, akokaram}@tcd.ie

Abstract. Temporal random variation of luminance in images can manifest in film and video due to a wide variety of sources. Typical in archived films, it also affects scenes recorded simultaneously with different cameras (e.g. for film special effect), and scenes affected by illumination problems. Many applications in Computer Vision and Image Processing that try to match images (e.g. for motion estimation, stereo vision, etc.) have to cope with this problem. The success of current techniques for dealing with this is limited by the non-linearity of severe distortion, the presence of motion and missing data (yielding outliers in the estimation process) and the lack of fast implementations in reconfigurable systems. This paper proposes a new process for stabilizing brightness fluctuations that improves the existing models. The article also introduces a new estimation method able to cope with outliers *in the joint distribution* of pairs images. The system implementation is based on the novel use of general purpose PC graphics hardware. The overall system presented here is able to deal with much more severe distortion than previously was the case, and in addition can operate at 7 fps on a 1.6GHz PC with broadcast standard definition images[1].

1 Introduction

Random fluctuation in the observed brightness of recorded image sequences, also called *flicker*, occur in a variety of situations, all well known in the vision and image processing community. The most commonly consumer observed instance of flicker is in archived film and video (see figure 1 and material [8]). It is caused by the degradation of the medium, varying exposure times of each frame of film, or curious effects of poor standards conversion. In any situation where multiple views are recorded by different cameras, the problem can also be observed due to different camera behaviour or lighting conditions due to camera orientation, even if video cameras have been previously calibrated using radiometric calibration routines. Both outdoor and indoor footage can show flicker due to

---

[1] This work has been founded by Enterprise Ireland, the EU project BRAVA, HEA PRTLI TRIP and the European project Prestospace.

Fig. 1. Example of flicker. See especially the black diagonal on the right frame

illumination variation. In addition, flicker can also affect modern film and video media especially when telecine transfer is not properly done.

Flicker is a global phenomenon, not spatially uniform, affecting the whole recorded image plane. As such its presence also has a detrimental effect on any application involving image matching since they tend to assume brightness constancy between frames, like in motion estimation [6, 4]. In addition, flicker reduces the redundancy between frames and hence increases the bandwidth of transmitted sequences. This is particularly a problem for Digital Television and the compression of multiview scenes. Dealing with the problem of flicker requires some attempt to model or measure the brightness fluctuation between frames (the estimation process), and then to remove this fluctuation in some way (the correction process). This paper presents a new robust technique to correct flicker in image sequences. By merging models previously proposed in the literature, we propose a new generic non-linear model to link two images affected by flicker. This non-linear transformation is then estimated using two approaches, one being the standard non-robust histogram matching technique [9]. As an alternative, we propose to make the estimation more robust using two methods. First, the non-linear function is directly estimated by Maximum A Posteriori on the joint distribution of the intensities of the two considered frames. Second, because this distribution is a mixture of data linked by the flicker model (inliers) and some mismatched data (outliers due to occlusions or motion), we propose an original inlier enhancement process to reduce the influence of outliers in the estimation. In addition, we present a novel implementation based on general purpose PC graphics hardware. Our results show that we are able to cope with a much wider range of flicker effect than previously was the case. We also present results showing the effect of flicker on MPEG4 compression of different kinds of sequences including multi-view camera sequences, and the improvement in bandwidth usage with our proposed flicker reduction method.

2   Related Works

The following paragraphs present several approaches that have been proposed in the literature - most of them independently - for the removal of flicker in videos. Three main models have been proposed for intensity distortion. All try to recover the unknown original intensity image (ground truth) I_n^o given the observed intensity images I_n at time n. The estimation $\hat{I}_n^o$ of the non-degraded image gives the restored frame I_n^R.

Linear Model with Spatial Dependencies. Decenciere [1] proposed a linear model, involving a gain a_n and an offset b_n, linking the intensities of the observed image I_n to the original flicker free image I_n^o. As illustrated in figure 1, the flicker is not the same across an entire frame. Roosmalen [12, 13] introduced then the spatial dependency (x, y) as follows:

$$I_n^o(x, y) = a_n(x, y)\, I_n(x, y) + b_n(x, y) \tag{1}$$

While processing the video, the original flicker free image $I_n^o(x, y)$ is not yet available at time n and is replaced in equation 1 by the previous estimated one $\hat{I}_{n-1}^o$. In Roosmalen's approach, spatial variation is introduced on a block basis, thus a_n, b_n are piecewise constant. Parameters are estimated by least squares estimation [14]. Each estimate is associated with a confidence measure [12].

Some pairs of blocks can not be matched because of missing data occurring for instance, with blotches, occlusions, etc. To cope with this problem, Roosmalen [12] and Yang [14] suggest to detect occluding areas based on spotting large intensity differences that cannot be explained by flicker alone. Parameter estimation is then performed only on the blocks in which there are no outliers detected. Estimates for the "missing blocks" are then generated by some suitable interpolation algorithm. Unfortunately, this method for detecting outliers fails in the presence of heavy flicker degradation.

Because the restored images are generated by *locking* the brightness to previous frames, errors can accumulate. To avoid this problem, due partly to a bias in the estimation, the final restored intensity image is given by a mixture of $\hat{I}_n^o$ and the observed image :

$$I_n^R(x, y) = k\, \hat{I}_n^o(x, y) + (1 - k)\, I_n(x, y) \tag{2}$$

where k is the forgetting factor set experimentally at 0.85 [13].

Linear Model with Robust Regression. Ohuchi [10] and also Lai [6] consider the same linear model as in equation (1) but the spatial variation of the gain and the offset is expressed directly using second order polynomials [6, 10]. The parameters of those polynomials are then estimated using robust M-estimation [6, 10, 3, 11] involving a Reweighted Least Squares algorithm. Robustness of the estimator is needed to deal with outliers that frequently occur in old videos.

However, as noticed by Kokaram et al.[5], due to the correlation between the frames, the regression (robust and non-robust) introduces a bias in the estimates that can damage seriously the restoration process in case of heavily degraded sequences. Therefore, the authors [5] have introduced a slightly modified linear model that allows reduction of this bias. Another improvement proposed in [5] is to change the polynomial basis to a cosine basis to express the gain and the offset. Since the success of global motion estimation is linked to flicker correction (and vice versa), some have proposed to couple both estimations [6, 13, 5]. In this paper, images will be registered prior to any flicker estimation.

Non-linear Model. However, the brightness distortion can sometimes be non-linear. Naranjo et al. [9] proposed the following non-linear model: $I_n(x, y) = f_n(I_n^o(x, y))$, where f_n is any increasing function estimated by comparing the intensity histogram of the current observed image and the average histogram of neighbouring frames [9] (see equation 5). Notably, this model does not account for the spatial variation of the flicker defect.

3   A New Model

In our experience with this problem, spatial dependence and non-linearity are key to modelling a wide range of flicker defects. Since the causes of flicker are usually unknown and various, we prefer to adopt a very weak prior on the distortion function. We propose therefore to extend the non-linear model proposed by Naranjo [9] to integrate spatial variations: $I_n^o(\mathbf{x}) = f_n(I_n(\mathbf{x}), \mathbf{x})$, where $\mathbf{x} = (x, y)$ is the pixel location, I^o and I the flicker free and the observed frames respectively.

Ideally, an estimate of a transformation should be performed for each pixel $\mathbf{x}$. However, this is computationally expensive and it would require an additive spatial smoothness constraint. Alternatively, we propose to estimate the transformations f_n^i at regularly spaced control points $\mathbf{x}_i$ in the image using interpolating splines of order 3. The splines yield and implicit smoothness constraint and our corrected pixel at $\mathbf{x}$ can be written: $I_n^o(\mathbf{x}) = \sum_i^N w(\mathbf{x} - \mathbf{x}_i) f_n^i(I_n(\mathbf{x}))$, where f_n^i is the transformation at control point $\mathbf{x}_i$ and $w(\mathbf{x})$ the interpolating 2D mask. The number of control points on one axis is defined as the *flicker order*.

Temporal Integration. A key deviation from previous efforts in flicker removal is that we model the change in brightness between observed images and we do not model the brightness change between the observed and the hidden clean images. Thus f is estimated between frames n, m and our new model is:

$$I_m(\mathbf{x}) = \sum_i^N w(\mathbf{x} - \mathbf{x}_i) f_{n,m}^i(I_n(\mathbf{x})) \tag{3}$$

Therefore our task is to smooth brightness changes between frames and not necessarily reveal the true original image. Brightness variations between images I_m and I_n can be caused by: 1) intentional effects like shadows or gradual brightness changes, due to special editing effects for instance, and 2) the flicker degradation which is unintentional. The first effect is generally low frequency and exhibits slow temporal variation. The second effect is temporally impulsive and it is this signal that has to be removed.

As illustrated in figure 2, we consider all the estimations $f_{n,m}$, between I_n and its neighbouring frames $\{I_m\}$. These estimated parameters $f_{n,m}$ correspond to the impulsive flicker mixed with the gradual informative variations. To

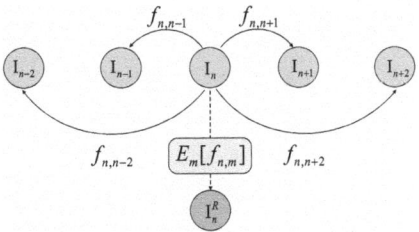

Fig. 2. Compensation of a frame: $\{I_n\}_n$ original pictures, I_n^R : restoration of I_n, $f_{n,m}$ brightness variation parameters between frames I_n and I_m

separate the impulsive flicker and keep the informative variation, a temporal robust expectation is computed:

$$\hat{f}_n(k) = \mathbb{E}_m\left[f_{n,m}(k)\right] = \arg\min_{f_n(k)} \sum_{m=n-M}^{n+M} e^{-\frac{(m-n)^2}{\sigma_w}} \rho(f_{n,m}(k) - f_n(k)) \qquad (4)$$

where $f_{n,m}(k)$ is the kth component of the lookup table $f_{n,m}$, ρ a robust function [11] and σ_w a temporal scale factor. The temporal window has been fixed experimentally to 15 frames ($M = 7$). The expectation $\hat{f}_n$ is estimated using Reweighted Least Squares and finally applied to I_n to generate the restored image I_n^R.

Speeding Up the Correction Scheme. It is possible to dramatically reduce the complexity by estimating $f_{n,m}$ through the simple combination of successive estimations $f_{n,n+1}$. However, when the flicker is too severe that only a portion of the intensity range is occupied by an image (skewed pdf) the parameter estimates are poor. To conclude, two strategies are available: one involving exhaustive estimations but able to cope with extreme fluctuations in intensity without propagation of errors and one involving a minimal number of estimations but more liable to errors.

4   Estimation of the Brightness Distortion Function $f_{n,m}^i$

In the case of a spatially constant brightness distortion ($I_m = f_{n,m}(I_n)$), it has been shown in [9] that the distortion function can be estimated by [2]:

$$f_{n,m}(I_n) = C_m^{-1} \circ C_n(I_n) \qquad (5)$$

where C_n and C_m are the cumulative histograms of I_n and I_m. In our case we can assume the distortion locally constant. At each control point $\mathbf{x}^i$, the local pdf is given by[2]:

$$p_n^i(u) = \frac{\sum_{\mathbf{x}|I_n(\mathbf{x})=u} w(\mathbf{x} - \mathbf{x}_i)}{\sum_{\mathbf{x}} w(\mathbf{x} - \mathbf{x}_i)} = \sum_{\mathbf{x}|I_n(\mathbf{x})=u} w(\mathbf{x} - \mathbf{x}_i) \qquad (6)$$

---

[2] Where $\sum_{\mathbf{x}} w(\mathbf{x}) = 1$ for splines.

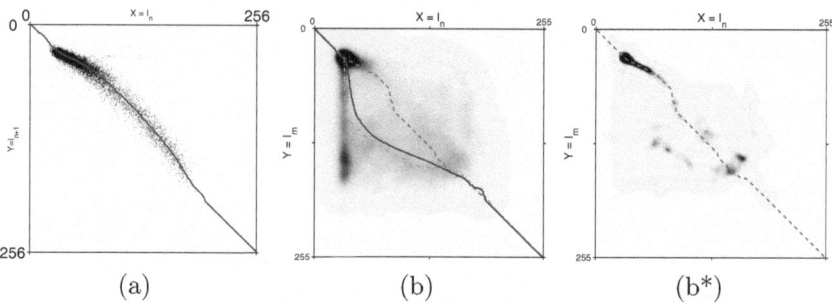

Fig. 3. Examples of joint pdf and their corresponding estimated distortion functions. On (a) an example of non linear distortion and its pdf matching estimation; on (b) an example of blotch noticeable by the vertical line and its consequence on the estimation by pdf matching (solid blue), by using MAP (dotted green) and by using the inlier enhancer and MAP (dashed red); on (b*): the joint pdf of (b) after inlier enhancement

where $\mathbf{x} \rightarrow w(\mathbf{x})$ is the spatial weighting function and $\mathbf{x}_i$ a control point. $\hat{f}^i_{n,m}$ is estimated using equation (5).

This *Pdf matching* scheme (eq. 5) is a non-biased estimator, unlike the regression methods [5]. Another interesting point about considering pdfs separately on each frames is that pdfs are not affected by local motion within the area of consideration, whereas regression methods assume that there is no local motion between both frames. Nevertheless, as shown in the next paragraph, this estimator is very sensitive to large occlusions. We present next an original method that allows us to cope with severe levels of occlusion.

A New Treatment of Outliers. Previous methods for outlier detection [12, 14] are based on brightness differences between frames. The detection is taken by thresholding the mean absolute error. However, in case of strong flicker, the differences might be only due to the flicker and not to the outliers. We propose here a new way to dissociate outliers and inliers based on the observation of the local joint pdf of the image intensities (similarly defined as in eq. 6). To simplify notation we denote U and V the random variables I_n and I_m and we write f instead of $f^i_{n,m}$.

Estimation of f by Maximum A Posteriori. We propose here to estimate f that maximises of the number of couples (u, v) that follow the flicker model (i.e. $f(u) = v$). In other words we want to maximise the likelihood $\prod_u \mathcal{P}(u, v = f(u)|f)$ with the prior knowledge that f is increasing[3]. This can be simply done in a Markovian framework by considering $f = [f(0), \ldots, f(u)]_{u \in 0:255}$ as a Markov chain of order 1, whose transition probabilities $\mathcal{P}(f(u + 1)|f(u))$ are such as

---

[3] For simplicity, we use $\mathcal{P}(u, v)$ instead of $\mathcal{P}_{UV}(U \in \mathcal{C}_u, V \in \mathcal{C}_v)$ and $\mathcal{P}(u)$ instead of $\mathcal{P}_U(U \in \mathcal{C}_u)$. The probabilities are in practice approximated by histograms.

to ensure $f(u+1) > f(u)$. It is possible as well to add boundary conditions on f so that $f(0) = 0$ and $f(255) = 255$.

$$\hat{f} = \arg\max_f \prod_{u=1}^{255} \mathcal{P}(u, f(u)|f) \cdot \mathcal{P}(f(u)|f(u-1)) \tag{7}$$

The Maximum A Posteriori $\hat{f}$ can be found by using the Viterbi algorithm [7].

Inliers Enhancement. However the MAP estimation can perform poorly in the presence of important occlusions.Figure 3-b shows the joint distribution of intensity between two images I_n and I_{n+1} corresponding to axis u and v respectively. If the images presented no outliers and were only degraded by brightness fluctuation (ie. no occlusions, no local motion), the distribution would be along the plot $v = f(u)$ (i.e. the inlier class $\mathcal{I}$). The case illustrated shows a region of missing data in frame $n + 1$ (typically a large region of constant color). This outlier causes a spurious ridge in the distribution manifesting as an horizontal line $u = u_0$. Sadly some of the probabilities on this outlier line are greater than the corresponding inliers one (i.e. $\mathcal{P}(u_0, f(u)) > \mathcal{P}(u, f(u))$) and the MAP estimation will try to follow this outlier line.

The task is to successfully reject the outlier ridge. Simply weighting out parts that are too far away from the identity line $u = v$ is the classical method used in robust regressions [5] but it becomes inefficient when the transformation f makes inlier pairs deviate a lot. To suppress the ridge, it would be intuitively beneficial to explore a weight $\mathcal{R}(u, v)$ on the joint distribution which make inliers row and column wise maximum:

$$\text{if } (u, v) \in \mathcal{I} \text{ then } \begin{cases} \forall u' \neq u, & \mathcal{R}(u, v) > \mathcal{R}(u', v) \\ \forall v' \neq v, & \mathcal{R}(u, v) > \mathcal{R}(u, v') \end{cases} \quad \text{where } f(u) = v \tag{8}$$

We propose to examine the factor :

$$\mathcal{R}(u, v) = \mathcal{P}(u|v)\, \mathcal{P}(v|u) = \frac{\mathcal{P}(u, v)^2}{\mathcal{P}(u)\, \mathcal{P}(v)} \tag{9}$$

Under some hypotheses (cf. Appendix A), it can be shown that the factor $\mathcal{R}$ fulfills conditions (8) and can therefore be used iteratively to enhance the distribution of the inliers in the mixture while reducing the outliers. Initialising $\mathcal{P}^{(0)}(u, v)$ at $\mathcal{P}(u, v)$, we iterate:

$$\mathcal{P}^{(n+1)}(u, v) = K^{(n+1)}\, \mathcal{P}(u, v)\, \mathcal{R}^{(n)}(u, v) \tag{10}$$

where $K^{(n+1)}$ is a normalising constant. It is shown experimentally that applying this procedure for $n = 3$ produces sufficient attenuation of the outlier distribution and then to improve the estimation of f (cf. section 6). Figure 3-b* shows some results obtained with this iterative process. We can see that the occlusion in sequence (b) has been efficiently removed. The estimation of f using MAP on $\mathcal{P}^{(n)}(u, v)$ is then improved.

5   Practical Issues

Tuning the Restoration Process Parameters. The flicker order (i.e. the number of control points per dimension) typically vary from 3—for the vast majority of image sequences—to 6 or even 14 for old movies whose film has chemically changed. For flicker orders greater than 6, occlusions (blotches and local motions) force us to use the robust scheme presented in paragraph 2. Working on color sequences can be done by simply stabilizing the luminance component of the frames.

Flicker Compensation Using Graphics Hardware. Computational load can be a key issue in a restoration process involving thousands of frames. To try and alleviate this, we focused on the flicker compensation stage of our algorithm. We found that on average flicker compensation accounted for 80% of the total restoration time for our non-robust restoration process.Modern computer graphics card are becoming much more programmable and contain powerful graphic processing units (GPUs). In fact they can now be considered as useful co-processors to the CPU. Flicker compensation on the GPU is only possible because of the latest advances in graphics hardware architecture especially full support for floating point accuracy. Using fragment programs we can perform the necessary operations to map f_n^i on an image block and multiply it by $w(\mathbf{x} - \mathbf{x}_i)$. We use vertex programs to correctly position the interpolating kernels on the image. Render-to-texture and floating point data are required for the summation of $f_n^i(\mathbf{I}(\mathbf{x}))\, w(\mathbf{x} - \mathbf{x}_i)$.

Performance. Figure 4 shows the results of using the GPU compared to the CPU. These results were obtained on a 1.6GHz Pentium 4 machine running Windows 2000, with an Nvidia GeForce FX5600 graphics card. Using the GPU implementation we reduced the time taken for flicker compensaton from 80% of the total restoration time to 55%. On average the GPU implementation is 3.5 times faster than the CPU implementation. Altogether the full non-robust scheme processes on average 6.8 frames per second at flicker order 3. The full robust scheme takes 1.5 frames per second whereas Roosmalen's process takes around 2 to 3 seconds a frame on similar hardware. For heavily degraded sequences the full process can take up to 20 seconds per frame.

6   Results

Sequences. The first sequence, *Snake*, composed of images from different uncalibrated cameras. This is a modern sequence actually used for post production.The second one, *Paula*, is a 8mm movie from a private repository, captured by pointing a DV camera at the projection screen; the asynchrony of the frame rates is the principle source of flicker. We also processed the *Tunnel* sequence (obtained by telecine) [13] and eventually we processed sequences from *Rory O'More*, a severely degraded movie from 1911.

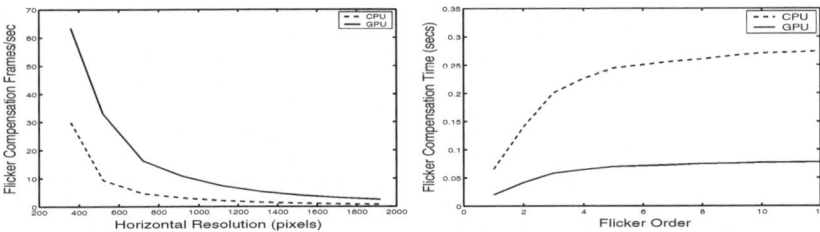

Fig. 4. Flicker Compensation CPU Vs GPU

Scenarios. Different restoration scenarios have been explored : no restoration at all (ob), naranjo's restoration (na), our method with *pdf matching* (pm), with MAP only(v), with MAP and inlier enhancement (hv) and finally our correction scheme assuming f linear and estimated as specified in [5] (af).

Evaluation by Comparing the Mean and the Variance. Assessing the performance of the systems on real degraded sequences is difficult because of the lack of objective criteria for assessing the quality of the restoration. However, as shown in [13, 10, 9, 5] it is feasible to expect that a good de-flicker process would reduce the fluctuations in the mean and the variance of image intensities from frame to frame. Figure 5-b shows some results for (pm) on the *Paula* and *Snake* sequences. We can see how the filter smoothed the brightness fluctuations. Figure 5-a shows the importance of the non-linear treatment. Whereas (af) cannot remove completely the fluctuations, (pm) stabilizes the brightness. However (na) hardly corrects the flicker.

Evaluation by Comparing the MPEG4 Compression Performances. However the mean and the variance cannot characterize subtle differences between restorations, especially if the dirts and blotches make the mean and variance fluctuate. We propose therefore a novel criteria for assessing the quality of flicker reduced, by comparing the compression ratio given by a MPEG4 encoder (in our case the Microsoft MPEG4 encoder). Results on the *Tunnel* sequence corroborate our previous remarks : (pm) improves the compression performances by 48.6%, (af) by 45.8% and (na) by only 38.4%. This criteria is also very useful to assess the improvements of our inlier enhancement heuristic (hv) as clearly shown in table 1 for damaged sequences from *Rory*.

However, as for the mean and the variance measure, this evaluation is still biased because it favours restoration processes that reduce details level.

7   Conclusion

Images sequences can be affected by brightness fluctuation for many reasons. We presented in this paper a new model able to deal with various kinds of flicker, a new process of correction, and a new efficient way for removal of outliers in the

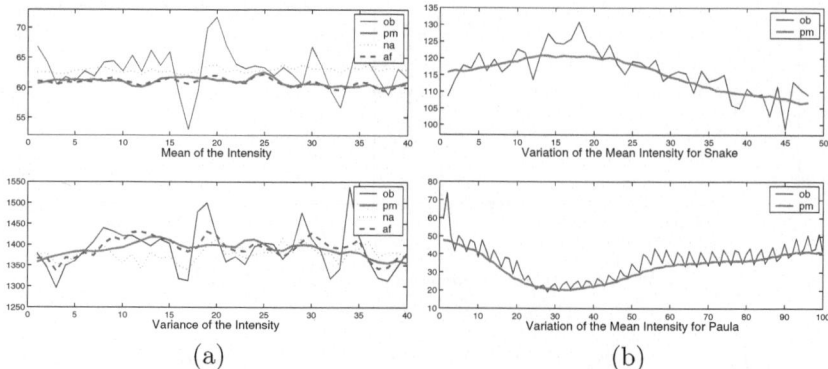

Fig. 5. Comparison for different scenarios of the mean and variance (a) (*Tunnel*). Variations of the mean (b)(top: *Snake*, bottom: *Paula*)

Table 1. Compressions ratios between the original compressed sequences and the restored ones. Better restorations are obtained for higher compressions ratios

Sequence	pm	v	hv
Rory (shot 11)	17%	9.9%	20.5%
Rory (shot 13)	12.4%	10.7%	17.2%
Rory (shot 16)	15.2%	8.0%	17.7%
Rory (shot 19)	4.3%	8.0%	9.0%
Tunnel	48.6%	-	45.6%
Paula	2%	-	12%

estimation of our brightness distortion function. These algorithms are fast and coupled with the use of cheap graphics hardware, we can reach near real-time performance on standard computers for most real image sequences.

A   The Factor $R(u,v)$

We want to show in a first case that $\mathcal{R}$ can actually hold condition (8) without any precondition on the outliers distribution. The only hypothesis made here is that the *proportion* of inliers for each color in both images is greater than the proportion of outliers: if $(u,v) \in \mathcal{I}$, we have $\mathcal{P}(u|v) > \frac{1}{2}$ and $\mathcal{P}(v|u) > \frac{1}{2}$.

$$\text{thus,} \quad \begin{cases} \text{if}(u,v) \in \mathcal{O}: \mathcal{P}(u|v) < \frac{1}{2} \\ \text{if}(u,v) \in \mathcal{O}: \mathcal{P}(v|u) < \frac{1}{2} \end{cases} \Rightarrow \begin{cases} \text{if}(u,v) \in \mathcal{I}: \mathcal{R}(u,v) > \frac{1}{4} \\ \text{if}(u,v) \in \mathcal{O}: \mathcal{R}(u,v) < \frac{1}{4} \end{cases} \quad (11)$$

The condition (8) is clearly fulfilled. But as the assumptions made on the amount of inliers is very strong, we would like to examine another case where a color might be occluded by more than 50% of outliers. As the case illustrated by figure 6-a is a very frequent one, we consider below its case study. In this problem, occlusions, all of color u_0, are only present in image I_n. Therefore the only non null probabilities are situated in the inlier parts $\mathcal{P}(u, f(u)) \geq 0$ and in the outliers parts corresponding to the monocolor occlusions $\mathcal{P}(u_0, .) \geq 0$. Thus if we introduce an inlier pair $(u_1, v_1) \neq (u_0, v_0)$ with $v_1 = f(u_1)$ and $v_0 = f(u_0)$, for condition (8) to hold, we need only:

$$\mathcal{R}(u_0, v_0) > \mathcal{R}(u_0, v_1) , \quad \mathcal{R}(u_1, v_1) > \mathcal{R}(u_0, v_1) \tag{12}$$

Demonstration: Let α be the proportion of occluded pixels. If we assume that the colour distribution of the occluded region ('under' the blotches) is the same as the colour distribution in unoccluded regions, we have:

$$\mathcal{P}(u_0, v_1) = \alpha \mathcal{P}(v_1) \Rightarrow \mathcal{P}(u_0|v_1) = \alpha \tag{13}$$

However, as occluded regions are usually spatially coherent, the intensity distributions will be different and we must consider a weaker hypothesis:

$$(H): \quad \alpha \leq M \leq 1 \mid \quad \mathcal{P}(u_0, v_1) < M \cdot \mathcal{P}(v_1) \quad ; \quad \mathcal{P}(u_0|v_1) < M \tag{14}$$

Note that this hypothesis means when $M > 1/2$ that for one intensity v_1 in image I_n, the amount of outliers can be greater than the amount of inliers (ie. $\mathcal{P}(u_0|v_1) > .5$). Now, from (H) we derive directly:

$$\mathcal{P}(u_0, v_0) = \mathcal{P}(v_0) = \frac{\mathcal{P}(v_0)}{\mathcal{P}(v_1)}\mathcal{P}(v_1) > \frac{\mathcal{P}(v_0)}{\mathcal{P}(v_1)}\frac{\mathcal{P}(u_0, v_1)}{M} \tag{15}$$

$$\mathcal{P}(v_1|u_0) < \frac{M}{K}\mathcal{P}(v_0|u_0) \quad \text{with} \quad K = \frac{\mathcal{P}(v_0)}{\mathcal{P}(v_1)} \tag{16}$$

$$\mathcal{P}(v_0|u_0) + \mathcal{P}(v_1|u_0) \leq 1 \quad \Rightarrow \quad \mathcal{P}(u_0|v_0) < \frac{1}{1 + M/K} \tag{17}$$

$$\mathcal{R}(u_0, v_0) = \mathcal{P}(u_0|v_0)\, \mathcal{P}(v_0|u_0) = \mathcal{P}(v_0|u_0) < \frac{1}{1 + M/K} \tag{18}$$

$$\mathcal{R}(u_1, v_0) = \mathcal{P}(u_1|v_0)\, \mathcal{P}(v_0|u_1) < \frac{M^2}{K}\mathcal{P}(v_0|u_0) = \frac{M^2}{K}\mathcal{R}(u_0, v_0) \quad \text{cf. } H, 16 \tag{19}$$

$$\mathcal{R}(u_1, v_1) = \mathcal{P}(u_1|v_1)\, \mathcal{P}(v_1|u_1) = \mathcal{P}(u_1|v_1) > 1 - M > \frac{1 - M}{1 + M/K}\mathcal{R}(u_0, v_0) \tag{20}$$

Finally to fulfill conditions (12), we need $\frac{M^2}{K} < 1$ and $\frac{1-M}{1+M/K} > \frac{M^2}{K}$. Solutions for K and M are shown on figure 6-b. Stated briefly, the more frequent the occlusion's corresponding color (v_0) in the picture $n + 1$, the better the algorithm works. In particular, if the occlusion color appears in the most frequent color ($K > 1$), the algorithm will work even if the full picture is half occluded ($M > .5$). In practice of course, the assumptions made here do not necessarily apply exactly. We find that this means that inliers could be slightly attenuated. However, because this process is applied together with the MAP estimation discussed in section 2, flicker is properly suppressed in all examples tested so far as shown in the material[8].

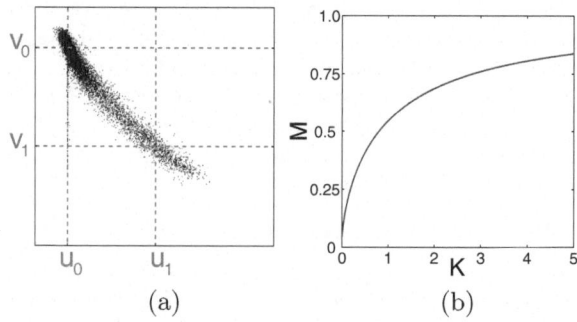

Fig. 6. On (a), example of single color occlusion. On (b) $K = \frac{\mathcal{P}(v_0)}{\mathcal{P}(v_1)}$ versus M (in case of uniform occlusion, M corresponds to the proportion of occluded pixels)

References

1. E. Decencière. *Restauration automatique de films anciens.* PhD thesis, Ecole Nationale Supérieure des Mines de Paris, December 1997.
2. R. Gonzalez and P. Wintz. *Digital Image Processing.* 2nd edition, 1987.
3. P.J. Huber. *Robust Statistics.* John Wiley and Sons, 1981.
4. H. Jin, P. Favaro, and S. Soatto. Real-Time feature tracking and outlier rejection with changes in illumination. In *ICCV*, pages 684–689, July 2001.
5. A. C. Kokaram, R. Dahyot, F. Pitie, and H. Denman. Simultaneous luminance and position stabilization for film and video. In *Visual Communications and Image Processing*, San Jose, California USA, January 2003.
6. S.-H. Lai and M. Fang. Robust and efficient image alignement. In *Proc. of the IEEE Conference on Computer Vision and Pattern Recognition (CVPR)*, volume 2, pages 167–172, Fort Collins, Colorado, June 1999.
7. B.H. Juang L.R. Rabiner. An introduction to hidden markov models. *IEEE ASSP Mag.*, pages 4–16, 1986.
8. Video Material. URL: http://papabois.mee.tcd.ie/sigmedia/publications/.
9. V. Naranjo and A. Albiol. Flicker reduction in old films. In *Proc. of the 2000 International Conference on Image Processing (ICIP-2000)*, September 2000.
10. T. Ohuchi, T. Seto, T. Komatsu, and T. Saito. A robust method of image flicker correction for heavily-corrupted old film sequences. In *Proc. of the 2000 International Conference on Image Processing (ICIP-2000)*, September 2000.
11. W. H. Press, S. A. Teukolsky, W. T. Vetterling, and B. P. Flannery. *Numerical Recipes in C - The Art of Scientific Computing.* Cambridge University Press, 1995.
12. P.M.B. van Roosmalen. *Restoration of archived film and video.* PhD thesis, Delft University of Technology, October 1999.
13. P.M.B. van Roosmalen, R.L. Lagendijk, and J. Biemond. Correction of intensity flicker in old film sequences. *IEEE Transactions on Circuits and Systems for Video Technology*, 9(7):1013–1019, 1999.
14. X. Yang and N. Chong. Enhanced approach to film flicker removal. *Proceedings of SPIE Applications of Digital Image Processing XXIII*, 4115:39–47, 2000.

Factorization of Natural 4 × 4 Patch Distributions

Kostadin Koroutchev and José R. Dorronsoro*

Depto. de Ingeniería Informática and Instituto de Ingeniería del Conocimiento,
Universidad Autónoma de Madrid, 28049 Madrid, Spain

Abstract. The lack of sufficient machine readable images makes impossible the direct computation of natural image 4 × 4 block statistics and one has to resort to indirect approximated methods to reduce their domain space. A natural approach to this is to collect statistics over compressed images; if the reconstruction quality is good enough, these statistics will be sufficiently representative. However, a requirement for easier statistics collection is that the method used provides a uniform representation of the compression information across all patches, something for which codebook techniques are well suited. We shall follow this approach here, using a fractal compression–inspired quantization scheme to approximate a given patch B by a triplet (D_B, μ_B, σ_B) with σ_B the patch's contrast, μ_B its brightness and D_B a codebook approximation to the mean–variance normalization $(B - \mu_B)/\sigma_B$ of B. The resulting reduction of the domain space makes feasible the computation of entropy and mutual information estimates that, in turn, suggest a factorization of the approximation of $p(B) \simeq p(D_B, \mu_B, \sigma_B)$ as $p(D_B, \mu_B, \sigma_B) \simeq p(D_B)p(\mu)p(\sigma)\Phi(\|\nabla B\|)$, with Φ being a high contrast correction.

1   Introduction

The importance for a wide range of topics, that go from standard visual information processing tasks to the study of basic human visual behavior, of understanding the statistical behavior of natural images is plainly obvious. However, direct statistics computation is not possible even for 4 × 4 natural image blocks: current lossles image compression techniques do not allow to go below 2.5 bits per pixel rates [12], which implies that the representation of 4 × 4 blocks will require in average about $16 \times 2.5 = 40$ bits. In other words, direct natural block statistics would require about $2^{40-16} \simeq 16 \times 10^6$ natural 1024×1024 images and, simply, there are not so many machine readable raw images. This has led many researchers to collect and analyze statistics not directly on blocks B but rather on appropriate, low dimensional transformations $T(B)$. Typically, the block transformation computed prior to statistics collection either reduces a

---

* With partial support of Spain's CICyT, TIC 01–572.

D. Comaniciu et al. (Eds.): SMVP 2004, LNCS 3247, pp. 165–174, 2004.

block's dimension by selecting a few points, projecting the block pixels on some directions or computing a certain integral transform [7, 10] or, on the other hand, $T(B)$ allows a certain reconstruction of B [4, 9]. An example of this are those wavelet transform methods that compute statistics over the first few wavelet components. This is approximately equivalent to compute statistics over lower resolutions of the original patches. A good recent review of the literature on the statistics of natural images is given in [11]. In our context, desirable characteristics of the $T(B)$ transformation are

1. The information loss due to $T(B)$ should be as small as possible, so that it is meaningful to deduce statistical properties of B from those of $T(B)$. Clearly this does not hold for the methods in the first class above.
2. The data structure of $T(B)$ should be obviously the same for all B so that statistics are collected over a uniform data set.
3. Finally, the computation of $T(B)$ should be quite fast, if only to alleviate the computationally demanding task of statistics collection.

The first requirement shows that methods of the second type are clearly the most appropriate ones, as they connect statistics collection with image compression, while the second requirement may penalize optimal compression schemes such as DCT JPEG or wavelets, as the resulting $T(B)$ could be highly non uniform. The natural alternative, that we shall use here, are codebook image compression methods that compress a given block B by choosing another D_B such that

$$D_B = \arg \min_{D \in \mathcal{D}} dist(B, D) = \arg \min_{D \in \mathcal{D}} \{||B - (\sigma_B D + \mu_B)||_\infty\}, \qquad (1)$$

with $\mathcal{D} = \{D\}$ a certain codebook of mean and variance normalized domains. Here $|| \cdot ||_\infty$ denotes the pixel–wise supremum norm and $\sigma_B D + \mu_B$ is a gray level transformation of D, with the block's standard deviation σ_B and mean μ_B being respectively taken as a contrast factor and a luminance shifting. B is then compressed by the triplet (D_B, σ_B, μ_B). We shall see this triplet as the transformation of B, that is, $T(B) = (D_B, \sigma_B, \mu_B)$. Although clearly inspired by fractal image compression, we shall look at the (D_B, σ_B, μ_B) coding and its associated reconstruction

$$B \simeq \sigma_B D_B + \mu_B \qquad (2)$$

from a codebook point view. In any case, the coding $T(B)$ certainly meets our second requirement and, as we shall see, also the first one, as the approximation it provides is close enough. Turning our attention to the third requirement, a fast computation of σ_B and μ_B can be easily achieved. Finding D_B, however, can very time consuming as it will involve full block comparisons. To minimize their number, we shall use here a hash based block pre–comparison, inspired in hash–based fractal image compression (FIC), a novel image compression method proposed by the authors [3], whose performance is comparable to other state of the art FIC methods (or even better in some instances). More precisely, we shall compute first a certain hash–like function $h(D)$ for all codebook domains, and distribute in the same linked list those D with the same h value. To code a

Fig. 1. Lena´s image is a well known source of fractal codebooks, but statistics computed from other codebooks are similar, provided the source image is "rich" enough, as the one from the Van Hateren's database exemplified here. Its decimated square center has been used as an alternative domain source

given block B, a set $H(B)$ of small perturbations of the hash value $h(B)$ will be computed and the full block comparisons in (1) will be done only between B and those D such that $h(D) \in H(B)$.

More details on this hash based block–domain matching are given in section 2, where we shall also discuss the basic statistics collection procedure for the $T(B) = (D_B, \sigma_B, \mu_B)$ approximations. In section 3 we shall compute the mutual information between the joint probability $p(D_B, \sigma_B, \mu_B)$ and the marginal probabilities $P(D_B)$, $p(\sigma_B)$ and $p(\mu_B)$ and see that, in a first approximation, we have for 4×4 natural image patches B that

$$p(D_B, \sigma_B, \mu_B) \simeq p(D_B)p(\mu_B)p(\sigma_B), \qquad (3)$$

while a second order approximation is

$$p(B) \simeq p(D_B)p(\mu_B)(p(\sigma_B)\Phi(||\nabla B||), \qquad (4)$$

with $\Phi(||\nabla B||)$ a high contrast correction. For this we shall approximate about 280 million natural 4×4 patches extracted from the well known van Hateren database [2] using two different FIC codebooks, derived from the well known Lena image and from a typical van Hateren image depicted in figure 1. In section 4 we shall also analyze the structure of the $p(D)$ and $p(\sigma)$ probabilities ($p(\mu)$ can be easily manipulated and does not carry significant information). We shall show that $p(\sigma)$ (that is independent from the codebook used) has an exponential structure and that $p(D_B)$ follows for both codebooks a nearly uniform behavior with respect to volume in image space. The paper ends with some other comments and pointers to further work.

Table 1. Different entropy measures (in bits) of image statistics using the Lena (second column) and van Hateren (third column) codebooks and limit estimates (fourth column) for them

Quantity	Value(Lena)	Value(VH)	Limit estimates
N	231511046	231441592	–
$\log_2 N$	27.7865	27.7861	–
$H(i, j, s, \sigma, \mu)$	26.7141	26.6517	29.84
$H(i, j, s)$	17.8156	17.6565	17.73
$I(i, j, s \| \sigma, \mu)$	1.5642	1.4674	0.427
$I/H(i, j, s, \sigma, \mu)$	5.86%	5.51%	1.43%
$H(\sigma, \mu)$	10.4627	10.4626	10.46
$I(\sigma \| \mu)$	0.1698	0.1697	0.115
$I(\sigma, \mu)/H(\sigma, \mu)$	1.62%	1.62%	1.10%

2   Methods

The representation $B = (\tilde{B}, \sigma_B, \mu_B) \simeq (D_B, \sigma_B, \mu_B) = T(B)$ implies that $\tilde{B} = (B - \mu_B)/\sigma_B \simeq D_B$. This approximation must hold for all block pixels which, as we shall argue below, suggests to define a hash–like function

$$h(D) = \sum_{h=1}^{H} \left(\left\lfloor \frac{D_{i_h j_h}}{\lambda} \right\rfloor \% C + \frac{C}{2} \right) C^{h-1} = \sum_{h=1}^{H} b_h C^{h-1}. \tag{5}$$

to speed up domain searches. Here we shall take $H = 5$ and the points $D_{i_h j_h}$, $1 \le h \le H$, used are the four corner pixels and an extra middle pixel. C will be 16 and the modulus operator $\%C$ gives integer values between $-C/2$ and $C/2$. Finally λ is chosen so that (5) defines an approximately uniform base C expansion that speeds up hash searches. Therefore, we want the b_h to be approximately uniformly distributed between 0 and 16, which can be achieved if λ is chosen so that the $D_{i_h j_h}$ are uniformly distributed in $[-\lambda\frac{C}{2}, \lambda\frac{C}{2}]$. An optimal λ would then be about 0.2, although we shall take $\lambda = 2$ in what follows. Codebook domains will be derived from a 256×256 versions of the Lena and van Hateren images by extracting all its 4×4 (overlapping) blocks. This gives $(256 - 4 + 1)^2 \simeq 2^{16}$ codebook domains, that become 2^{20} after adding for each block its 8 isometries and its negative (notice that the dilations in (2) are positive). Flat domains, i.e., those such that $\sigma(D) \le 4$, may give distorted values in (5) and we will exclude them (about 25% of both codebook domains).

Full block comparisons for a natural block B, that is, the computation of $dist(B, D) = \sup |B_{ij} - \sigma_B D_{ij} - \mu_B|$ in (1) over all block pixels, are performed only over domains D such that $h(D) \in H(B)$, with $H(B) = \{h_\delta(B)\}$, where

$$h_\delta(B) = \sum_{h=1}^{H} \left(\left\lfloor \frac{B_{i_h j_h} - \mu_B}{\lambda \sigma_B} + \delta_h \right\rfloor \% C + \frac{C}{2} \right) C^{h-1} = \sum_{h=1}^{H} r_h^\delta C^{h-1}, \tag{6}$$

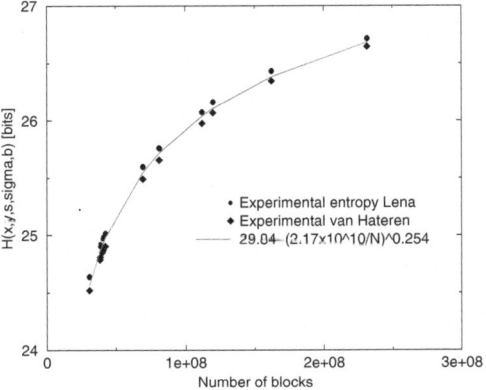

Fig. 2. Large sample behavior of the total entropy $H(i, j, s, \sigma, \mu)$ estimates. computed over Lena and van Hateren codebooks. Both values are close, but do not reach a saturation limit

with the displacement vector $\delta = (\delta_1, \ldots, \delta_H)^t$ verifying $|\delta_h| \leq 1$. Notice that the equality $h_\delta(B) = h(D)$ implies pixel closeness of the H points used to define h and, hence, a starting similarity between the D and B blocks. As a further acceleration factor, we shall content ourselves with approximate domain searches, in the sense that we will fix a tolerance value d and stop looking for domains matching a patch B as soon as a D is found such that $d(B, D) \leq d$. We shall take $d = 8$, that guarantees a reasonable reconstruction PSNR of about 30 Dbs. Finally, the coding of B will then be

$$T(B) = (i, j, s, \sigma, \mu)$$

where (i, j) indicates the position in the codebook image of the left upper corner of the matching domain, and s is an index for the isometry and negative used. As the natural patch source, we shall work with 4300 8 bit gray level images of size 1540×1024 from the van Hateren database. We shall restrict ourselves to their 1024×1024 squared centers. As done for domains, we will also exclude flat blocks, that is, those B with $\sigma \leq 4$ (about 20% of all database patches). This leaves us with a sample of about 232×10^6 natural 4×4 patches.

3   Distribution Factorization

Denoting by N the number of sample patches and by M the number of (i, j, s, σ, μ) probability bins, we should have $N \gg M$ in order to achieve accurate entropy estimates [8]. However, table 1 shows this not to be the case when estimating the full sample entropy $H_N(i, j, s, \sigma, \mu)$ of the $p(i, j, s, \sigma, \mu)$ distribution, something that can also be appreciated in figure 2, that shows that although close, the Lena and van Hateren full entropy values do not reach a saturation

 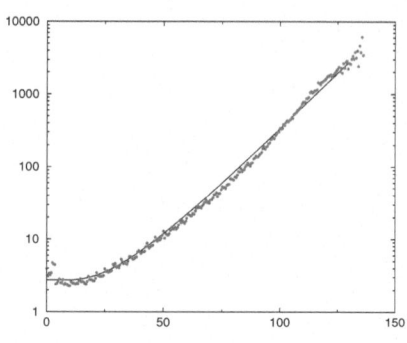

Fig. 3. Left: deviations from the first order factorization (7) arise over the edges of the codebook source image. Right: values of $E_{||\nabla B||}$ in logarithmic scale and second order epitaxy inspired approximation

point. On the other hand, we have indeed $N \gg M$ when estimating the marginal entropies $H_N(i,j)$, $H_N(\sigma)$ and $H_N(\mu)$. As it can be seen in table 1, the Lena and van Hateren $H_N(i,j)$ entropy values are again very close ($H_N(\sigma)$ and $H_N(\mu)$ are codebook–independent). From the table one can deduce that σ and μ can be taken to be independent, as their mutual information is less than 2% of their joint entropy. When looking at the dependence between the (i,j,s) and (σ,μ) distributions, the table shows that the mutual information $I(i,j,s||\sigma,\mu)$ is for both codebooks about 1.5 bits, that is, about 5.5% of the joint entropy. Although not totally independent, this points out to a first order factorization of the joint (i,j,s,σ,μ) density as the product

$$p(i,j,s,\sigma,\mu) \simeq p(i,j,s)p(\sigma)p(\mu). \tag{7}$$

In order to visualize where the remaining 1.5 bit dependence may arise, we have looked at the average $d(i,j)$ over (s,σ,μ) of the quotient

$$\frac{p(i,j,s,\sigma,\mu)}{p(i,j,s)p(\sigma)p(\mu)}.$$

The values of $d(i,j)$ can be projected over the codebook source image, where we should look for those different from one. When this is done (see figure 3, left, where $\log d(i,j)$ is depicted), it is clear that image edges are where to look in order to correct (7). This also suggests to correct (7) as

$$p(B) \simeq p(D_B)p(\mu_B)p(\sigma_B)\Phi(||\nabla B||).$$

This second approximation clearly implies that we should have

$$\Phi(||\nabla B||) \simeq E_{||\nabla B||}\left[\frac{p(i,j,s,\sigma,\mu)}{(p(i,j,s)p(\sigma)p(\mu)}\right] \tag{8}$$

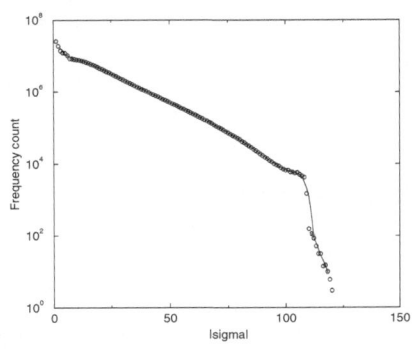

Fig. 4. Left: projection over Lena of the $-\log p(\sigma(i,j))$ distribution. Right: relative frequencies of σ in a logarithmic scale. The linear central behavior suggests an exponential distribution

with $E_{||\nabla B||}$ denoting the conditional expectation with respect to $||\nabla B||$. This expectation is depicted in figure 3, right, which also shows that a good approximation for $\Phi(x)$ is given by

$$\Phi(x) \simeq a + b\phi\left(\frac{x}{c}\right) = a + b\left(\frac{x}{c} - \frac{\pi^2}{6} - 2\sum_1^\infty \frac{(-1)^n}{n^2} e^{-\frac{x}{c}n^2}\right),$$

where $a \simeq 2$, $c \simeq 31$ and $b/c \simeq 0.068$. The motivation for the $\phi(t)$ function comes from epitaxy studies in crystal growth [5], where it models the time evolution of the number of nuclei in non steady state crystal nucleation. Once the Φ–dependence is taken into account, the mutual information between the experimentally obtained distribution and the above second order corrected distribution is now 0.621 bits, that is, about 1 bit less than the previous estimate. Therefore, just 2.3% of the total information is not covered now by the second approximation.

4   Structure of the Codebook and Contrast Distributions

The histogram of the σ distribution is depicted in figure 4. It has a drop around 100, due to the limited range of brightness levels, and also a cusp–like peak at 0, mostly due to the many near flat blocks that arise from the layered structure of natural images. In any case, its structure is carried by the central linear zone, that suggests an exponential distribution. Moreover, and as it could be expected, $p(\sigma)$ is quite correlated with the codebook source's edges, as depicted in figure 4, left, that shows for the Lena image the values of $-\log p(\sigma(i,j))$ with $\sigma(i,j)$ the σ value for the domain with i, j as its upper left corner coordinates.

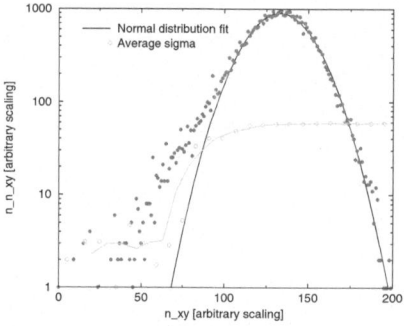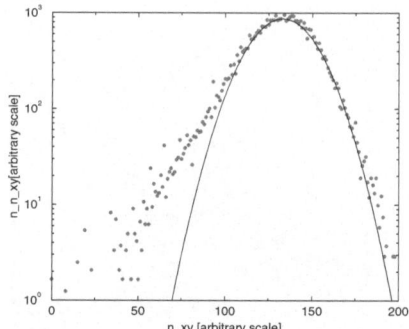

Fig. 5. Volume–corrected values of $\log N(m)$, with $N(m)$ the relative number of domains getting m natural blocks for the Lena (left) and van Hateren (right) codebooks. The distributions are fairly similar and suggest a volume–uniform distribution of all normalized natural patches among codebook domains. The left image also shows the average patch σ as a growing function of m. The scale of σ goes for a low value of about 5 at left, to a high value of 13.4 at the right high count area

To get a hold on the $p(i,j)$ distribution, we may count for each m' the number $N'(m')$ of domains that match exactly m' patches. In other words, we look at the (i,j) as a bins where matching patches fall. The resulting distribution has a clear parabolic structure, that suggests that domains "fall" more or less uniformly on bins or, more precisely, that in block space, patches are more or less distributed uniformly among domains. However, this uniform distribution hypothesis requires also to take into account the volume surrounding each domain when counting $N'(m')$, for then the probability of a codebook region $\mathcal{R}$ receiving a patch is proportional to its volume $v(\mathcal{R})$. This would lead to the rather difficult problem of estimating this surrounding volume $v(D)$ for each domain. To avoid it, we have made the simplifying assumption that all hash linked lists correspond to domain space regions of the same volume and, therefore, to estimate $v(D)$ as a multiple of $1/\nu(h)$, with $\nu(h)$ the number of domains in the list indexed by $h = h(D)$. We should therefore correct the basic count m' of patches matched by a domain D to $m = m' \times \nu(h(D))$. The corrected $N(m)$ values are depicted in figure 5 in logarithmic scale, for the Lena (left) and van Hateren (right) codebooks. Both show a very similar parabolic structure, in which the left side divergence is due to low σ patches, with small denominators in (6) and, hence, more sensible to noise variations that may alter the matching domain they are assigned to. Notice that this effect should be more marked in domains getting fewer blocks. This is supported in figure 5, left, that shows for m the average σ value among the $N(m)$ patches: it is about 5 for small m but goes up to 13.4 for m in the high count area.

It is clear that figure 5 is best explained as a large sample gaussian aproximation of a binomial distribution or, in other words, that the codebook indices (i,j) do follow a volume–uniform distribution. Apparently this may contradict

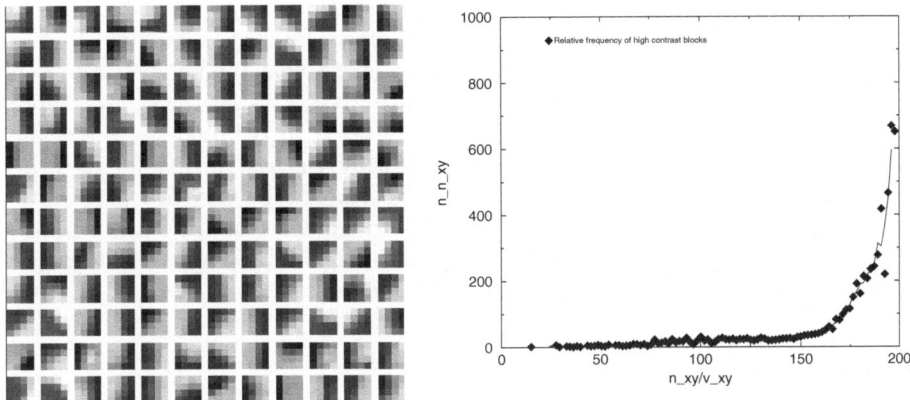

Fig. 6. Left: domains with a higher count of high contrast patches. They clearly correspond to edges. Right: the proportion of high contrast patches is markedly higher among high count domains

recent results in [4], that show a marked structure for high contrast natural 3×3 blocks. However, notice that figure 6, right, depicting the proportion of high-contrast patches as a function of m, has a sharp rise at the high patch count area. Moreover, when the domains receiving the largest count of high contrast patches are depicted, as in figure 6, left, it is clear that they correspond to edges.

Finally, we just mention that the μ distribution is highly dependent on factors such as the camera's calibration and can be easily manipulated through, say, histogram equalization. Thus, it does not carry significant information.

5   Summary and Future Directions

To alleviate the large dimension of the state space of 4×4 natural patches, we have proposed in this work to estimate their distributions in terms of an image compression inspired codebook approximation of the form $B \simeq (D_B, \sigma, \mu)$, with σ, μ the block's variance and mean and D_B a codebook domain close to the normalization of B. Identifying a domain D_B in terms of its (i, j) location on the source image and the symmetry s applied, we have also shown how to factorize the distribution $p(i, j, s, \sigma, \mu)$ as $p(i, j, s, \sigma, \mu) \simeq p(i, j, s)p(\sigma)p(\mu)\Phi(||\nabla B||)$. Of these factors, the most relevant in terms of information seems to be $p(\sigma)\Phi(||\nabla B||)$ combination, which allows us to conclude that, at least in the scale investigated (about one minute of angle), the information is essentially carried by the block's edges. This is certainly not surprising, as it agrees with the well known Marr hypothesis [6]. However, this conclusion is achieved here through direct information theoretic considerations; in particular, they are independent of any consideration regarding the receiving system. In turn, this could suggest that biological systems have adapted themselves to extract those natural image parts most relevant in terms of information theory.

Moreover, the structure of the marginal distributions $p(i, j)$ and $p(\sigma)$ may have practical applications in areas such as image database searching. In fact, the absence of long tails in these distributions shows that the patches' representation proposed here has a very compact range. Thus, using the (i, j, σ) representation as a key for database searching, the worst distributed key is actually the exponentially distributed image contrast, while the search time for the other distributions should be nearly constant. Furthermore, the very fast drop of the exponential distribution makes it reasonable to expect that the codebook coding scheme proposed here should allow for fast image database search strategies. This and other related topics are currently being studied.

References

1. Y. Fisher (ed.), **Fractal Image Compression: Theory and Application**, Springer Verlag, New York, 1995.
2. J.H. van Hateren and A. van der Schaaf, *Independent component filters of natural images compared with simple cells in primary visual cortex.* Proc.R.Soc.Lond. B, 265 (1998), 359-366.
3. K. Koroutchev and J. Dorronsoro, *Hash–like Fractal Image Compression with Linear Execution Time*, Iberian Conference on Pattern Recognition and Image Analysis, IbPRIA 2003. Lecture Notes in Computer Science 2652 (2003), 395–402.
4. A.B. Lee, K.S. Pedersen and D. Mumford. *The Complex Statistics of High-Contrast Patches in Natural Images.* WWW Proceedings of Second International IEEE Workshop on Statistical and Computational Theories of Vision. Vancouver, Canada, July 2001.
5. I.M. Markov, **Crystal growth for beginners: Fundamentals of Nucleation, Crystal Growth and Epitaxy**, World Scientific, 1995.
6. D. Marr, **Vision,** W.H. Freeman and Co., 1982.
7. B.A. Olshausen and D.J. Field, *Natural image statistics and efficient coding*, Workshop on Information Theory and the Brain, Network: Computation in Neural Systems 7 (1996), 336–339.
8. L. Paninski, *Estimation of Entropy and Mutual Information*, Neural Computation 15 (2003), 1191–1253.
9. K.S. Pedersen and A.B. Lee. *Toward a Full Probability Model of Edges in Natural Images.* Proceedings of the 7th European Conference on Computer Vision, Copenhagen, Denmark. Lecture Notes in Computer Science 2350 (2002), 328–342.
10. D.L. Ruderman, *The statistics of natural images*, Network: Computation in Neural Systems 5 (1994), 517-548.
11. A. Srivastava, A.B. Lee, E.P. Simoncelli and S.C. Zhu, *On Advances in Statistical Modeling of Natural Images*, Journal of Mathematical Imaging and Vision 18 (2003), 17-33.
12. M. Weinberger, G. Seroussi and G. Sapiro, *LOCO-I: A Low Complexity, Context-Based, Lossless Image Compression Algorithm*, Proc. of the IEEE Data Compression Conference, Snowbird, Utah, March-April 1996.

Parametric and Non-parametric Methods for Linear Extraction

Benedicte Bascle, Xiang Gao, and Visvanathan Ramesh

Siemens Corporate Research,
755 College Road East, Princeton, NJ 08540, USA

Abstract. This article presents two new approaches, one parametric and one non-parametric, to the linear grouping of image features. They are based on the Bayesian Hough Transform, which takes into account feature uncertainty. Our main contribution are two new ways to detect the most significant modes of the Hough Transform. Traditionally, this is done by non-maximum suppression. However, in truth, Hough bins measure the likelihoods not of single lines but of collection of lines. Therefore finding lines by non-maxima suppression is not appropriate. This article presents two alternatives. The first method uses bin integration, automatic pruning and fusion to perform mode detection. The second approach detects dominant modes using variable bandwidth mean shift. The advantages of these algorithms are that: (1) the uncertainties associated with feature measurements are taken into account during voting and mode estimation (2) dominant modes are detected in ways that are more correct and less sensitive to errors and biases than non-maxima suppression. The methods can be used with any feature type and any associated feature detection algorithm provided that it outputs a feature position, orientation and covariance matrices. Results illustrate the approaches.

1 Introduction

Feature detection and grouping are some of the basic functions of computer vision. Among possible approaches for feature detection and grouping, the Hough Transform (HT) and its extensions are a well-known method for extracting geometric curves [1]. In particular the linear HT is useful in man-made environments where linear features are abundant. It can thus be applied in many applications, such as detection, recognition, etc. There is an abundant literature on the Hough Transform. Numerous derived approaches (probabilistic HT, randomized HT, etc.) exist [2] [3] [4] [5] [6].

The performance of the Hough Transform is sensitive to a couple of issues. First, image noise causes errors in the estimation of feature position and orientation. This makes votes in Hough space noisy and perturbs their accumulation. A solution to this is the Bayesian Hough Transform [7](BHT). In the BHT, each feature casts weighted votes in several bins, according to a Gaussian

D. Comaniciu et al. (Eds.): SMVP 2004, LNCS 3247, pp. 175–186, 2004.
© Springer-Verlag Berlin Heidelberg 2004

distribution with an estimated mean and covariance matrix. Taking such covariance into account gives the BHT some robustness against noise. A second issue affecting HT performance is the detection of significant lines in the Hough accumulator. Traditionally, local maxima are detected. However the localization of local maxima is dependent on the distribution of bins in Hough space. Therefore line estimation by non-maxima suppression can be biased. In this article, we present two variants of the Bayesian Hough Transform that do not use non-maxima suppression to detect the most significant modes in Hough space. The first approach uses a pruning and fusion mechanism to find lines in the Hough accumulator, while the second approach is based on variable bandwidth mean shift.

Section 2 gives a short overview of feature detectors that can be used as input to the Hough Transform. In particular, their statistical properties are summarized. Section 3 presents the Bayesian Hough Transform (BHT) introduced by [7], from which our methods are derived. Section 4 describes our first method for line detection, the BHT with integration and pruning. Section 5 shows our second method, the BHT with 3D Bayesian Hough space and variable bandwidth mean shift. Finally, section 6 shows experimental results.

2   Feature Detection with Covariance Matrices

Traditionally, the Hough Transform module accepts any types of feature as input, as long as that feature can be described by its pose, orientation and the variance of the orientation. The variance of the feature position is generally assumed to be 1 to a few pixels. The variance of the feature orientation depends on the filter used to estimate feature orientation.

Examples of calculation of feature orientation variance are shown in table 1 for any linear convolution kernel (second column of the table) [8] [9] and for the facet estimation kernel (third column of the table, see [10] for a description of the facet kernel). The table shows how the variance σ_I^2 of image noise can be estimated from the least-square residual between the original image I and a filtered image J (after linear filtering/smoothing or intensity surface approximation). Note that σ_I represents not only true image noise but also the goodness of fit between our image model and the true image. In particular, σ_I can be chosen to encompass the variance of image texture and shading in the neighborhood of features. In this way, such texture and shading are modeled as normal variations in image intensity, not to be detected as features. Once the image noise is estimated, the variance σ_θ^2 of the orientation θ of a detected feature can be estimated. For instance, for an edge point detected by linear filtering and gradient estimation, the orientation is: $\theta = \arctan \frac{g_y}{g_x}$ and the variance of the orientation is: $\sigma_\theta^2 = \frac{\sigma_J^2}{||g||^2}$. Table 1 shows the corresponding variance estimate for facet detection.

Table 1. Feature detection modules and characterization of their performance

	Image linear filtering	facet estimation
	smooth image	linear facet
Image Model	$J = W * I$ with W any linear filter (prewitt, sobel, Gaussian approximation, etc.)	$I = \alpha * x + \beta * y + \gamma$ with $\alpha = \frac{\sum_x \sum_y x I(x,y)}{\sum_x \sum_y x^2}$ $\beta = \frac{\sum_x \sum_y y I(x,y)}{\sum_x \sum_y y^2}$ $\gamma = \frac{\sum_x \sum_y I(x,y)}{\sum_x \sum_y 1}$
Image Noise	$\sigma_I^2 = \sum_x \sum_y [J(x,y) - I(x,y)]^2$ $\sigma_J^2 = (\sum_i w_i^2)\sigma_I^2$	$\sigma_I^2 = \sum_x \sum_y [I(x,y) - \alpha y - \beta x - \gamma]^2$ $\sigma_\alpha^2 = \frac{\sigma_I^2}{\sum_y \sum_x x^2}$ $\sigma_\beta^2 = \frac{\sigma_I^2}{\sum_y \sum_x y^2}$ $\sigma_{\alpha\beta} = \frac{\sigma_I^2 \sum_y \sum_x xy}{\sum_y \sum_x x^2 \sum_y \sum_x y^2}$
Feature	edge point	edge point
Orientation	$\theta = \arctan \frac{g_y}{g_x}$	$\theta = \arctan \frac{\alpha}{\beta}$
Orientation Variance	$\sigma_\theta^2 = \frac{2\sigma_J^2}{\|g\|^2}$	$\sigma_\theta^2 = \left(\frac{\sigma_\alpha^2}{\beta^2} + \frac{\alpha^2 \sigma_\beta^2}{\beta^4} - \frac{2\alpha\sigma_{\alpha\beta}}{\beta^3} \right) \cos^4\theta$

3   State of the Art: Bayesian Hough Transform with Non-maxima Suppression

The Bayesian Hough Transform (BHT) has been introduced by Ji and Haralick [7]. The BHT has a probabilistic voting scheme which takes into account feature uncertainty. In the classic Hough Transform, a feature point (x, y) with orientation θ votes for one line $(\theta, \rho = x \cos \theta + y \sin \theta)$ in the accumulator space. In the BHT, the same edge point votes for a bundle of lines (θ', ρ') where $\theta' \in [\theta - 3\sigma_\theta, \theta + 3\sigma_\theta]$ and $\rho' \in [\rho - 3\sigma_\rho, \rho + 3\sigma_\rho]$. Note that this 3σ interval corresponds the 99.7% confidence interval of a 1D Gaussian distribution. The votes around (θ, ρ) are given by a normal distribution $\mathcal{N}_{[(\theta,\rho),\Sigma_{(\theta,\rho)}]}$ centered at (θ, ρ) and with covariance matrix $\Sigma_{(\theta,\rho)}$. This means that a vote $V_{(x,y)}(\theta', \rho')$ in a bin (θ', ρ') is equal to:

$$V_{(x,y)}(\theta', \rho') = \mathcal{N}_{[(\theta,\rho),\Sigma_{(\theta,\rho)}]}(\theta', \rho')$$

$$= \frac{1}{2\pi |\Sigma_{(\theta,\rho)}|^{\frac{1}{2}}} \exp \left\{ -\frac{1}{2} \begin{bmatrix} \theta' - \theta \\ \rho' - \rho \end{bmatrix}^T \Sigma_{(\theta,\rho)}^{-1} \begin{bmatrix} \theta' - \theta \\ \rho' - \rho \end{bmatrix} \right\}$$

The covariance matrix is a function of the feature orientation and 2D pose covariances σ_θ^2 and $\sigma_P^2 = \sigma_x^2 = \sigma_y^2$, as follows:

$$\Sigma_{(\theta,\rho)} = \begin{bmatrix} \sigma_\theta^2 & \sigma_{\theta\rho} \\ \sigma_{\theta\rho} & \sigma_\rho^2 \end{bmatrix} = \begin{bmatrix} \sigma_\theta^2 & k\sigma_\theta^2 \\ k\sigma_\theta^2 & k^2\sigma_\theta^2 + \sigma_P^2 \end{bmatrix} \text{ with } k = y\cos\theta - x\sin\theta \quad (1)$$

k is the distance from the origin to the line (θ, ρ) in image space. σ_θ^2 and $\sigma_P^2 = \sigma_x^2 = \sigma_y^2$ are functions of the feature detector used. Section 2 gives the values of σ_θ^2 for a few different feature detectors. In practice, $\sigma_P^2 = \sigma_x^2 = \sigma_y^2$ is often considered to be one to a few pixels. Note that, where image contrast is high, features pose and orientation usually have low covariance and thus Gaussian-distributed voting by the Bayesian Hough Transform voting is reduced to voting for one point, as does the classic HT.

Line hypotheses are generated in a standard way by finding local maxima of the Hough accumulator. However this approach can generate biases and multiple instances of equivalent line hypotheses. This is due to both image space and Hough space discretization. Image discretization means that edge point orientation estimation is biased. Hough space discretization means that, for voting purposes, the Hough space is partitioned into bins. The size and distribution of these bins has an impact on the distribution of local maxima in the Hough accumulator. For instance, in some cases, one single line in the image can give raise to two local maxima in Hough space, if voting for the line happens to fall around the border between two bins.

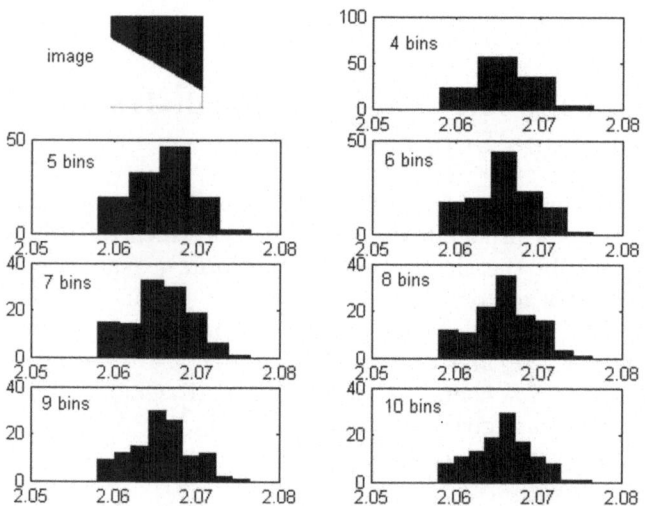

Fig. 1. Motivation of new approaches: finding the location of significant modes in Hough space by non-maxima suppression is subject to errors due to image and Hough space discretization. Here we see how the location of the local maximum is influenced by the number of bins used to calculate the edge point orientation histogram

These issues are illustrated by the example shown in figure 1. Let us consider a synthetic image of a step edge of orientation 30 degrees, integrated by a CCD camera with a point spread function of one. The image (with dynamic range $0 - 255$) is corrupted by Gaussian noise of standard deviation 2. After smoothing by a 5×5 Gaussian kernel of $\sigma = 1$, edge points are extracted and their orientations are calculated. The mean of the orientation distribution is biased by 1.6 degrees with respect to the real orientation. The figure shows the orientation histograms obtained using different number of bins (4 to 10). The position of the peak of the distribution depends on the bin distribution and can be biased by up to 1.3 degrees. On an image of width 512, this means that in some places along the lines, line points might be off with respect to image edge points by as much as 10 pixels. This example shows evidence that non-maxima suppression over bins can give biased estimates of image lines. To remedy this problem, we propose two new alternative methods to the Bayesian Hough Transform with non-maxima suppression.

4 Line Detection Using Bayesian Hough Transform with Evidence Integration and Pruning

This section presents our first approach to detecting lines by finding significant modes in Hough space without using non-maxima suppression. The approach differs from the Bayesian Hough Transform ([7]) by: (1) performing bin-size dependent integration (2) automatically pruning and fusioning line hypotheses using statistical priors (without any non-maxima suppression).

4.1 Bayesian Hough Transform with Integration (BHTi)

In the Bayesian Hough Transform proposed by Ji and Haralick ([7]) and summarized in section 3, a given feature point throws one weighted vote per bin. The amplitude of the vote is given by sampling the normal distribution associated with the vote at the center of the bin. This means that voting in two bins of the same center but different sizes would be identical. In order to make voting reflect bin partition, this paper proposes to integrate the bivariate normal distribution over the bin, instead of simply sampling it at the center of the bin. This evidence integration also makes sense in view of the fact that a single bin $[\rho' - \Delta\rho, \rho' + \Delta\rho] \times [\theta' - \Delta\theta, \theta' + \Delta\theta]$ represents the likelihood of a group of lines. We call this new variant the Bayesian Hough Transform with integration (BHTi). With doing this integration, the vote in a bin centered on (θ', ρ') from the voting feature point (x, y, θ) becomes (with $\rho = x\cos\theta + y\sin\theta$):

$$V_{(x,y)}(\theta', \rho') = \int_{\rho'-\Delta\rho}^{\rho'+\Delta\rho} \int_{\theta'-\Delta\theta}^{\theta'+\Delta\theta} \mathcal{N}_{[(\theta,\rho), \Sigma_{(\theta,\rho)}]}(\theta'', \rho'')d\rho''d\theta'' \qquad (2)$$

$$= \int_{\rho'-\Delta\rho}^{\rho'+\Delta\rho} \int_{\theta'-\Delta\theta}^{\theta'+\Delta\theta} \frac{1}{2\pi|\Sigma_{(\theta,\rho)}|^{\frac{1}{2}}} \exp\left\{ -\frac{1}{2} \begin{bmatrix} \theta'' - \theta \\ \rho'' - \rho \end{bmatrix}^T \Sigma_{(\theta,\rho)}^{-1} \begin{bmatrix} \theta'' - \theta \\ \rho'' - \rho \end{bmatrix} \right\}$$

$\Delta\theta$ and $\Delta\rho$ represent respectively the θ and ρ bin sizes in the Hough accumulator. The bivariate normal integral above can be approximated by doing a variable change to a coordinate frame where the covariance ellipse is aligned with the axes and centered on the origin. In this coordinate frame, the bivariate normal integral becomes separable and can be calculated from the tables of the cumulative distribution function of the standardized normal distribution.

4.2   Hypothesis Pruning by Automatic Hough Accumulator Thresholding

A threshold can be estimated automatically for the Bayesian Hough Transform with integration (BHTi), as follows. First, as shown by equation 1, the covariance matrix $\Sigma_{(\theta,\rho)}$ of the Hough vote cast by a single feature is only a function of k and σ_θ, where k is the distance from the voting feature point to the point on the line closest to the origin and σ_θ is the standard deviation of the feature orientation. In the case when edge point features are used in conjunction with the BHTi, σ_θ is a function of the intensity gradient norm $\|\nabla I\|$ only and thus the covariance matrix $\Sigma_{(\theta,\rho)}$ of a Hough vote is only a function of k and $\|\nabla I\|$. As described before, the Hough vote is cast probabilistically into several bins according to the Gaussian distribution given by $\Sigma_{(\theta,\rho)}$. The maximum contribution is cast into the bin containing (θ, ρ). The amplitude of this contribution $V_{\max}$ depends on the distance between (θ, ρ) and the bin center (θ', ρ') (see equation 2). A maximum value of

$$V_{\max} \text{ is: } \int_{-\Delta\rho}^{\Delta\rho} \int_{-\Delta\theta}^{\Delta\theta} \frac{1}{2\pi|\Sigma_{(\theta,\rho)}|^{\frac{1}{2}}} \exp\left\{-\frac{1}{2}\begin{bmatrix}\theta''' - \theta \\ \rho''' - \rho\end{bmatrix}^T \Sigma_{(\theta,\rho)}^{-1} \begin{bmatrix}\theta''' - \theta \\ \rho''' - \rho\end{bmatrix}\right\} d\rho''' d\theta'''.$$

Given that $\Sigma_{(\theta,\rho)}$ depends only on k and σ_θ (as discussed above), $V_{\max}$ is also a function of k and σ_θ. Given a-priori distributions for k and σ_θ, the average vote in a central bin is: $\bar{V}_{\text{feature}} = \frac{1}{C} \int_{\sigma_\theta} \int_k V_{\max}(k, \sigma_\theta) d\sigma_\theta dk$, with C a normalizing constant. In addition, the probability that a feature is present at a particular pixel in an image tile can be approximated by: p(a feature is present at pixel p in tile T) $= \frac{N_{\text{features}}}{M_T \cdot N_T}$, where N_{features} is the number of features found in the image tile and (M_T, N_T) are the dimensions of the image tile. Then the average vote cast by a pixel p in tile T is: $V_{\text{pixel } p \in \text{ tile } T} = \bar{V}_{\text{feature}} \cdot p$(a feature is present at pixel p in tile T). If the minimum a-priori segment length is $L_{\min}$, the vote accumulated from all the pixels on the segment is $V_{\min} = L_{\min} * V_{\text{pixel } p \in \text{ tile } T}$. If we model the vote generated by a segment as a binomial distribution, then its variance can be written as: $\sigma_V = \sqrt{L_{\min} V_{\text{pixel } p \in T}(1 - V_{\text{pixel } p \in T})}$. Then a choice of threshold is: $T_{HT} = V_{\min} + 2\sigma_V$ (which corresponds to the 95% confidence interval).

These calculations give us an automatic way of setting a threshold on the Hough Transform accumulator, based on the statistical properties of the processed images. To our knowledge, this is the first paper that describes an automatic approach to Bayesian Hough buffer thresholding. By contrast, [7] use an arbitrary threshold. For each bin where voting exceeds the threshold, a line hypothesis is generated. Thresholding is the only pruning mechanism used here. Non-maxima suppression is not used, contrary to most approaches in the literature. This allows line hypotheses to be generated even for lines that are close to local maxima of votes, but not local maxima themselves. This is a good idea because, as discussed in section 3, a local maximum in Hough space does not necessarily give the true parameters of a line but might give slightly biased parameters.

Additionally, if priors are available about line orientation and position, these are used to prune line hypotheses. And this process (hypothesis pruning) can also be applied in a multi-scale scheme.

4.3   Hypothesis Fusion and Verification

To deal with the potential generation of multiple hypothesis corresponding to the same image line, statistical hypothesis fusion is performed. Application priors about the lengths of segments and their distances to other segments are applied to do this. Surviving and fused line hypotheses are validated or discarded by projection and verification in image space.

5   Line Detection Using 3D Bayesian Hough Space and Variable Bandwidth Mean Shift

This section presents a second and different method for detecting lines by finding significant modes in Hough space without using non-maxima suppression (which is potentially erroneous). The main idea here is to use variable bandwidth mean shift to detect Hough modes. This also leads us to use a 3D Bayesian Hough space, instead of a 2D space as in most approaches in the literature, including [1] [2] [3] [4] [5] [6] [7].

5.1   Bayesian Hough Transform with 3D Hough Accumulator

The Bayesian Hough Transform proposed by Ji and Haralick [7] and described in section 3 uses a 2D Hough accumulator (ρ, θ). We propose here to use a 3D Hough accumulator $(\rho, \cos \theta, \sin \theta)$. The advantage of such an accumulator is that wraparound issues associated with $0 \leq \theta < 2\pi$ are avoided during the mean shift procedure. We also model the uncertainty in the estimate θ of feature orientation by a Von-Mises distribution with parameter θ_0 and κ. The Von Mises distribution is the circular analog of the normal distribution on a line. It is defined on the range $[0, 2\pi)$. The representation of the distribution of orientations due to noise by a VonMises distribution is more accurate than by a Gaussian distribution. Let θ_0 be the mean orientation measured at the feature point (x_0, y_0), we have

$$E \left\{ \begin{bmatrix} x \\ y \\ \theta \end{bmatrix} \right\} = \begin{bmatrix} x_0 \\ y_0 \\ \theta_0 \end{bmatrix} \tag{3}$$

And,

$$E \left\{ \begin{bmatrix} \rho \\ \cos \theta \\ \sin \theta \end{bmatrix} \right\} = \frac{I_1(\kappa)}{I_0(\kappa)} \begin{bmatrix} x_0 \cos \theta_0 + y_0 \sin \theta_0 \\ \cos \theta_0 \\ \sin \theta_0 \end{bmatrix} \tag{4}$$

with $\kappa = \frac{||g||^2}{\sigma_j^2}$ where $||g||$ is the norm of the gradient and σ_j^2 is the variance of the intensity in the smoothed image.

The covariance matrix of the vector $\begin{bmatrix} \rho \\ \cos\theta \\ \sin\theta \end{bmatrix}$ is $\begin{bmatrix} \sigma^2_\rho & \sigma^2_{\rho\cos\theta} & \sigma^2_{\rho\sin\theta} \\ \sigma^2_{\rho\cos\theta} & \sigma^2_{\cos\theta} & \sigma^2_{\sin\theta\cos\theta} \\ \sigma^2_{\rho\sin\theta} & \sigma^2_{\sin\theta\cos\theta} & \sigma^2_{\sin\theta} \end{bmatrix}$.

The equations below give us the details of the covariance matrix associated with each vote $(\rho, \cos\theta, \sin\theta)$ in 3D Hough space.

$$\sigma^2_{\cos\theta} = \frac{1}{2} + \frac{1}{2}\cos 2\theta_0 \frac{I_2(\kappa)}{I_0(\kappa)} - \cos^2\theta_0 \left(\frac{I_1(\kappa)}{I_0(\kappa)}\right)^2$$

$$\sigma^2_{\sin\theta\cos\theta} = \frac{1}{2}\sin 2\theta_0 \left[\frac{I_2(\kappa)}{I_0(\kappa)} - \left(\frac{I_1(\kappa)}{I_0(\kappa)}\right)^2\right]$$

$$\sigma^2_{\sin\theta} = \frac{1}{2} - \frac{1}{2}\cos 2\theta_0 \frac{I_2(\kappa)}{I_0(\kappa)} - \sin^2\theta_0 \left(\frac{I_1(\kappa)}{I_0(\kappa)}\right)^2$$

$$\sigma^2_{\rho\cos\theta} = x_0\sigma^2_{\cos\theta} + y_0\sigma^2_{\sin\theta\cos\theta}$$

$$\sigma^2_{\rho\sin\theta} = x_0\sigma^2_{\sin\theta\cos\theta} + y_0\sigma^2_{\sin\theta}$$

$$\sigma^2_\rho = \sigma^2_{xx}\left(\frac{1}{2} + \frac{1}{2}\cos 2\theta_0 \frac{I_2(\kappa)}{I_0(\kappa)}\right) + \sigma^2_{yy}\left(\frac{1}{2} - \frac{1}{2}\cos 2\theta_0 \frac{I_2(\kappa)}{I_0(\kappa)}\right)$$

$$+ \sigma^2_{xy}\sin 2\theta_0 \frac{I_2(\kappa)}{I_0(\kappa)} + x_0^2\sigma^2_{\cos\theta} + y_0^2\sigma^2_{\sin\theta} + 2x_0y_0\sigma^2_{\sin\theta\cos\theta}$$

where $\Delta x = x - x_0$ and $\Delta y = y - y_0$.

5.2   Mode Detection Using Variable Bandwidth Mean Shift

As discussed in section 3, line hypotheses are usually generated by finding local maxima of the Hough accumulator. However this is subject to errors and biases. Section 4 presented a first alternative approach, based on hypothesis pruning and fusion, to detecting lines in Hough space. Here we propose a second alternative approach. To exploit the uncertainty information of the votes, the idea is to extract modes corresponding to significant lines from the Hough accumulator using variable bandwidth mean shift (VBMS) [11] [12]. This is one of the new contributions of this paper.

The Hough accumulator is built as described in the previous section. And the variable bandwidth (VBMS) mean shift is applied to find dominant modes in the accumulator. VBMS is a mode-seeking procedure and proceeds as follows. Given n data points with their uncertainty $\{(x_i, C_i)\}_{i=1,\ldots,n}$ and a starting point $x^{(0)}$, the corresponding significant mode $\hat{x}^{(m)}$ is sought iteratively using the following equation:

$$\hat{x}^{(m+1)} = H_h(\hat{x}^{(m)}) \sum_{i=1}^{n} \omega_i(\hat{x}^{(m)}) H_i^{-1} x_i \tag{5}$$

$$H_h(\hat{x}^{(m)}) = \left(\sum_{i=1}^{n} \omega_i(\hat{x}^{(m)}) H_i^{-1}\right)^{-1} \tag{6}$$

where

$$\omega_i(\hat{x}^{(m)}) = \frac{|H_i|^{-\frac{1}{2}} \exp\{-\frac{1}{2}(x_i - \hat{x}^{(m)})^T H_i^{-1}(x_i - \hat{x}^{(m)})\}}{\sum_{i=1}^{n} |H_i|^{-\frac{1}{2}} \exp\{-\frac{1}{2}(x_i - \hat{x}^{(m)})^T H_i^{-1}(x_i - \hat{x}^{(m)})\}} \quad (7)$$

and $H_i = C_i + \alpha I$ where α is a bandwidth parameter.

All votes in Hough space are used as starting points. The partition of the Hough space is obtained by grouping together all the data points that converged to the same point. Though binning can be used to speed up mean shift, it can also be avoided if computation time is not a concern. In this case, the advantage of using mean shift is that it potentially avoids some of the errors introduced by binning Hough space.

6   Experiment Results

In this section, we present some experimental results of applying our two approaches to line detection in images. The features used are edge points.

Figures 2 and 5 and 4 show the results of applying our first line detection method, the Bayesian Hough Transform with integration and pruning (see section 4), to part inspection and parcel detection. Figure 2 shows images of cast metal parts taken over a conveyor belt for inspection purposes. Figure 5 shows postal parcels coming along a conveyor belt underneath the camera. Figure 4 shows the Bayesian hough space for a structured image. These examples demonstrate that (1) the method is accurate (2) segments of varying length (relatively short segments as well as long ones) are recovered.

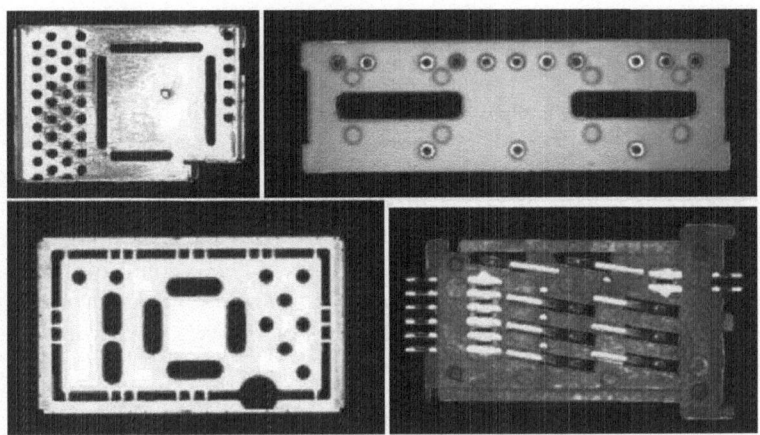

Fig. 2. Part inspection results using our first approach, the Bayesian Hough Transform with integration and pruning (see section 4). Extracted lines are overlayed in green

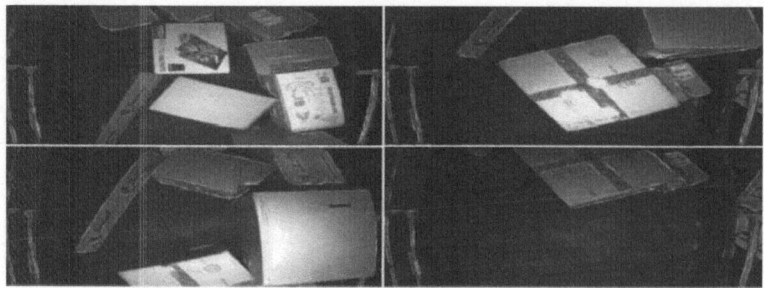

Fig. 3. Parcel singulation results using our first approach, the Bayesian Hough Transform with integration and pruning (see section 4). Extracted lines are overlayed in green

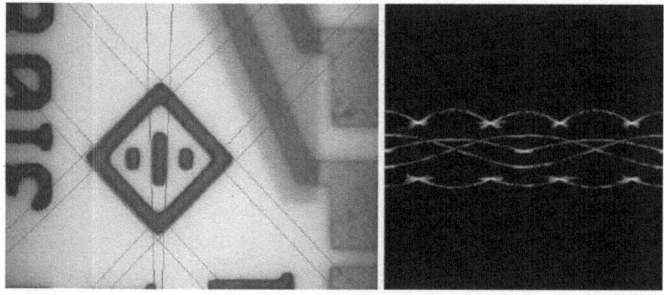

Fig. 4. Bayesian Hough space obtained using our first approach, the Bayesian Hough Transform with integration and pruning (see section 4)

Figure 5 show the results of applying our second line detection method, the 3D Bayesian Hough Transform with Variable Bandwidth Mean Shift (see section 5) to parcel singulation. Lines with limited support are more difficult to extract. We havent conducted a formal comparison of both approaches yet. However a few observations can be made about the two methods. The first one is parametric. It is more suited to a parallel SIMD implementation. It can detect short lines. It can also detect multiple lines that are very close to each other in an image. It is suited to difficult applications of the Hough Transform. The second approach is nonparametric, and less suited to detecting short segments. Future work includes a formal comparison between the performances of the two algorithms. A hybrid method is also possible.

7   Summary and Conclusions

This papers presents a system for the detection of linear structures in images of manmade objects. One of the characteristics of the system is its treatment of uncertainty from the bottom up. Two new methods are available for line extrac-

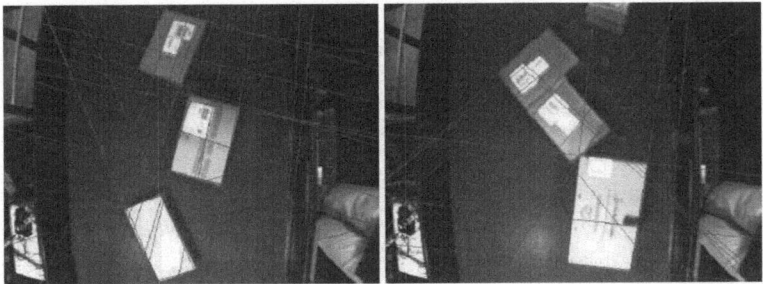

Fig. 5. Parcel singulation results using the 3D Bayesian Hough Transform with Variable Bandwidth Mean Shift (VBMS) (see section 5). Extracted lines are overlayed in red

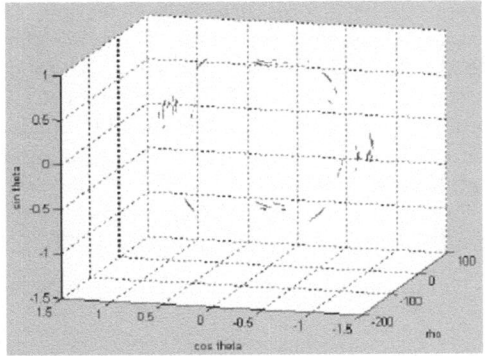

Fig. 6. Bayesian Hough space obtained using our second approach, the 3D Bayesian Hough Transform with Variable Bandwidth Mean Shift (VBMS), on a parcel image

tion. The first one, the Bayesian Hough Transform with integration and pruning, is more traditional, but uses statistics-based pruning to prune Hough hypotheses, instead of the classical non-maxima suppression. The second approach, the 3D Bayesian Hough Transform with Variable Bandwidth Mean Shift, performs mode detection using a 3D accumulator and variable bandwidth mean shift. Both approaches are motivated by detecting significant Hough modes in a way that is less prone to errors than non-maxima suppression over bins. Both methods show good performance. Future work includes a formal evaluation and comparison of the two algorithms.

References

1. Hough, P.: Method and means for recognition complex patterns. US patent 3,069,654 (1962)

2. Grimson, W., Huttenlocher, D.: On the sensitivity of the Hough transform for object recognition. IEEE Transactions on Pattern Analysis and Machine Intelligence **12** (1990) 255–274
3. Kalviainen, H., Hirvonen, P.: Connective randomized Hough transform (CRHT). Technical Report 44, Dept. of Information Technology, Lappeenranta University of Technology, Lappeenranta, Finland (1993)
4. Kalviainen, H., Hirvonen, P., Oja, E.: Houghtool – a software package for Hough transform calculation. In: Proceedings of the 9th Scandinavian Conference on Image Analysis, Uppsala, Sweden (1995) 841–848
5. Shaked, D., Yaron, O., Kiryati, N.: Deriving stopping rules for the probabilistic Hough transform by sequential analysis,. Computer Vision and Image Understanding **63** (1996) 512–526
6. McLaughlin, R., Alder, M.: The Hough transform versus the upwrite. IEEE Transactions on Pattern Analysis and Machine Intelligence **20** (1998) 396–400
7. Ji, Q., Haralick, R.: Error propagation for the Hough transform. Pattern Recognition Letters **22** (2001) 813–823
8. Canny, J.: A computational approach to edge detection. IEEE Transactions on Pattern Analysis and Machine Intelligence **8** (1986) 679–698
9. Gonzalez, R., Woods, R.: Digital image processing. Addison-Wesley (1992)
10. Haralick, R., Watson, L.: A facet model for image data. CGIP **15** (1981) 113–129
11. Comaniciu, D., Ramesh, V., Meer, P.: The variable bandwidth mean shift and data-driven scale selection. In: IEEE Int. Conf. Computer Vision. Volume 1., Vancouver, Canada (2001) 438–445
12. Comaniciu, D.: An algorithm for data-driven bandwidth selection. IEEE Transactions on Pattern Analysis and Machine Intelligence **25** (2003) 281–288

Crowd Segmentation Through Emergent Labeling

Peter H. Tu and Jens Rittscher

GE Global Research, One Research Circle, Niskayuna, NY 12309, USA
{peter.tu, jens.rittscher}@research.ge.com

Abstract. As an alternative to crowd segmentation using model-based object detection methods which depend on learned appearance models, we propose a paradigm that only makes use of low-level interest points. Here the detection of objects of interest is formulated as a clustering problem. The set of feature points are associated with vertices of a graph. Edges connect vertices based on the plausibility that the two vertices could have been generated from the same object. The task of object detection amounts to identifying a specific set of cliques of this graph. Since the topology of the graph is constrained by a geometric appearance model the maximal cliques can be enumerated directly. Each vertex of the graph can be a member of multiple maximal cliques. We need to find an assignment such that every vertex is only assigned to a single clique. An optimal assignment with respect to a global score function is estimated though a technique akin to soft-assign which can be viewed as a form of relaxation labeling that propagates constraints from regions of low to high ambiguity. No prior knowledge regarding the number of people in the scene is required.

1   Introduction

A crowd segmentation algorithm can be used to both update and initialize a tracking system. Typically the state of system is written as

$$X := (x_{00}, \ldots, x_{0N}, \ldots, x_{K0}, \ldots, x_{KN})^T \, , \tag{1}$$

where N denotes the number of model parameters and K the number of people in the scene. Initialization amounts to estimating the prior $P(\mathbf{X})$. The appearance of people varies in shape as well as attire. A number of approaches [2, 12] model the envelope of possible shapes based on silhouettes. The extraction of silhouettes of individuals in crowds itself is a very challenging problem. In this particular case occlusion is reduced by choosing a top view rather than an oblique perspective view. Assuming these imaging conditions one could argue that the crowd segmentation problem could be easily addressed by applying simple background subtraction followed by simple segmentation or standard blob forming techniques. Intille *et al.* [5] explored this approach and report that the blob detection is very unstable. One reason is that spurious background responses distort the blobs. In addition blobs merge whenever people stand close to each other and hence coalesce (see also figure 2 left). One key problem with many segmentation or blob detection approaches is that the number of people is not known in advance. Rather than making decisions solely on local measurements our approach uses a global criteria. Hence no prior knowledge regarding the number of people in the scene is required.

D. Comaniciu et al. (Eds.): SMVP 2004, LNCS 3247, pp. 187–198, 2004.

The work presented here formulates the segmentation of crowds into a set of individuals as a graph based grouping problem. Very basic shape information is incorporated by constraining the set of maximal cliques of the graph. In this work, reliance on well formed edge maps is avoided by restricting image measurements to a set of sparse interest points. The task of segmenting a crowd into a set of individuals can now be viewed as a model based grouping problem.

The set of interest points form the vertices $\mathbf{V}$. The existence of an edge e_{ij} between a pairs of vertices implies the plausibility that the two vertices could have been generated from the same individual. Low level image measurements are used to compute the weights a_{ij} of each of the edges, $e_{ij} \in \mathbf{E}$ in the graph $\mathbf{G}(\mathbf{V}, \mathbf{E})$. A newly introduced graph partitioning method, termed *emergent labeling* effectively incorporates geometric constraints based on a persons' shape into the graph structure. The algorithm used to compute an optimal partitioning function effectively propagates certainty from regions of low to high ambiguity. The partitioning function $\mathbf{L}$ directly maps to a particular state $\mathbf{X}$. It is argued that, rather than the partitioning function $\mathbf{L}$, the set of vertices $\mathbf{V}$ and associated edge weights are generated from a random process which depends on a set of parameters that control the feature extraction process. In order to model the uncertainly of associating an interest point with a foreground patch, an acceptance probability is introduced. The prior $P(\mathbf{X})$ is then computed by sampling different sets of vertices based on the acceptance probability. The algorithm does not need prior information such as the number of people in the site and it terminates in polynomial time.

1.1   Related Work

Very successful approaches were introduced in [9, 2] by directly learning the appearance of people using machine learning techniques. While support vector machines are applied in [9], Gravrila *et al.* [2] build up a hierarchical classifier for extracted silhouette shapes using a large amount of training data. Both of these approaches are well suited for detecting individuals which are separated from each other. Ronfard *et al.* [10] learn the appearance of body parts using support vector machines and then construct a likelihood function based on these detections.

[12] use a much simplified model for a human but make clever use of the scene geometry to locate people in a scene. The algorithm computes an initial set of head positions which subsequently serves as candidate points for a stochastic optimization. Their measurement process consists of computing edge maps, which can be very noisy under certain imaging conditions. Using constraints based on multiple views Mittal and Davis [8] segment a reasonably sized crowd into individuals. Here we restrict ourself to a monocular view but choose a top view rather than on oblique view of the scene.

1.2   Problem Statement

Image Measurement. The main goal of the measurement process is of course the identification of a set of interest points, $\mathbf{V} = \{v_i\}$ which can be reliably associated with the objects of interest, i.e. the people in the scene. In the first step, a probabilistic background model based on techniques presented in [1] is built. Image locations which indicate a high temporal and/or spatial discontinuity are then selected as feature points (see section 4 for details).

Graph Generation. Each feature point is associated with a vertex of a graph G. An edge e_{ij} between a pair of vertices v_i and v_j exists if and only if it is possible that the two vertices could have been generated from the same individual. This decision is entirely based on a geometric model of an average person's dimensions. A person's size with respect to image coordinates is determined using a combination of the homography between the image plane and the ground plane and the homography between the head plane and the image plane (see section 3). The strength of an edge e_{ij} is a function of the probability that the two connected vertices belong to the same individual. Unlike the edge existence criteria, the edge strengths are computed from observed image intensity values. A matrix A is defined such that A_{ij} is the strength of edge e_{ij}. A formal presentation of the graph generation process is given in section 4.

Graph Partitioning. Given the measurements embedded in the graph G the objective is to determine the true state of the system. The challenges associated with this task are twofold. First, the number of individuals in the scene is unknown. Second, if there is little separation between individuals, the computed inter-cluster edge strengths could be as strong as the intra-cluster edge strengths. Clustering algorithms such as k-means and normalized cut [11] have been applied to this type of problem. However the underlying assumption of these approaches is that the intra-cluster edge strengths of the true clusters are considerably stronger than the inter-cluster edge strengths. As previously stated, under crowded conditions this assumption is often violated. This problem can be overcome by effectively enforcing model constraints and propagating unambiguous assignments through the graph. A novel graph partitioning algorithm, referred to as *emergent labeling*, follows this principle and is presented in section 2.

Generating $P(X)$. For a given graph G the partitioning function L is derived deterministically. But the graph itself is, as already argued in the introduction, a result of a random process. In particular the uncertainly associated with the selection of the interest points needs to be accounted for. In section 6 an acceptance probability for an interest point will be derived. As it will become apparent, computing the strength of an edge e_{ij} is, for similar reasons, uncertain. By sampling from both the set of potential feature points and possible edge strengths associated with them, $P(X)$ can be computed.

2   Emergent Labeling

If a set of vertices constitutes a clique c, then there exists an edge between every pair of vertices in c. A maximal clique c on G is defined such that c is not a subset of any other clique on G. A clique is a very strict representation of a cluster. Because of this various authors have applied maximal clique based clustering algorithms to both affinity [6] and association graphs [7]. In the *emergent labeling* scheme, it is insisted that each vertex cluster in the estimate of the true state X be a clique on the graph G.

The set C is defined as all the maximal cliques on the graph G. In general computing C is an NP complete problem, however in our case the geometric shape model constrains the topology of G so that all the maximal cliques can be enumerated using direct methods. This process is discussed in section 3. Each vertex can be a member of multiple maximal cliques, the goal of the emergent labeling algorithm is to assign each vertex to only one of its maximal cliques. All the vertices assigned to the same maximal clique form a cluster in the estimate of the state $\mathbf{X}$. These clusters are by definition cliques although they are not necessarily maximal. This assignment process can be represented by a binary label matrix $\mathbf{L}$ which is defined as:

$$\mathbf{L}_{ij} = \begin{cases} 1 & \text{if } \mathbf{v}_i \text{ is assigned to } \mathbf{c}_j, \\ 0 & \text{otherwise.} \end{cases} \tag{2}$$

Since each vertex can be assigned to only one maximal clique, the sum of all the elements of each row of $\mathbf{L}$ must equal one.

Since under crowded conditions, it has been observed that it can be too confusing to make vertex assignment decisions based solely on local context, a global score function $S(\mathbf{L})$ is defined. By assuming that the true state $\mathbf{X}$ is optimal with respect to S, the vertex assignment decisions will be made on both local and global criteria.

2.1   Global Score Functional

A natural criteria for judging the merit of a cluster is to take the sum of the edge strengths connecting all the vertices inside that cluster. A global score functional can then be defined as the sum of the selected cluster strengths. By defining the affinity matrix $\mathbf{A}$ such that $\mathbf{a}_{ij}$ is equal to the edge strength of edge $\mathbf{e}_{ij}$ which has a value ranging between 0 and 1, the score function can be computed in the following manner:

$$S(\mathbf{L}) = \mathbf{trace}(\mathbf{L^t AL}). \tag{3}$$

The assignment matrix $\mathbf{L}$ defines a sub graph of $\mathbf{G}$ where all edges that connect vertices that have been assigned to different cliques are removed. The score function $S(\mathbf{L})$ is essentially the sum of the edge strengths in that sub graph. This score function tends to favor a small set of large cliques instead of a large number of small cliques. This can be seen by the fact that if a clique of size N were to be divided into two sub cliques of size N_1 and N_2 then there would be $N_1 \times N_2$ fewer edges contributing to the score function.

2.2   Optimization

Finding the optimal labeling matrix $\mathbf{L}$ with respect to the optimization criteria S is an integer programming problem which cannot be directly solved in an efficient manner. A soft assign approach similar to that developed in [3] will be used. In [3], the authors are attempting to solve a version of the assignment problem. Given two sets of entities, a valid mapping is defined such that each element in the first set can only be assigned to one element in the second set and no two elements in the first set can be assigned to the same element in the second set. Given the strengths associated with any pairwise assignment, the global score of a given mapping is the sum of the assignment strengths. The algorithm attempts to search for the mapping with highest global score.

In the version of soft assign presented in this paper, the first set of entities are the vertices of the graph and the second set of entities are the cliques. Like [3] each vertex can only be assigned to one clique, however they cannot be assigned to cliques to which they are not a member. Unlike [3] multiple vertex can be assigned to the same clique. Since there is no concept of a direct assignment strength between a vertex and a clique, the mapping score is set to the sum of the edge strengths between vertices that have been assigned to the same clique. Although the approaches share a similar formalism, they are radically different algorithms due to the differences in mapping restrictions and global score evaluation. Initially $\mathbf{L}$ is viewed as a continuous matrix so that each vertex can be associated with multiple cliques. After a number of iterations the matrix is forced to have only binary values.

For iteration $t + 1$, if vertex $\mathbf{v_i}$ is in clique $\mathbf{c_j}$ then the variable

$$r_{ij}(t+1) = e^{\beta \frac{dS(\mathbf{L}(t))}{d\mathbf{L_{ij}}}}. \tag{4}$$

The derivative is calculated by

$$\frac{dS(\mathbf{L}(t))}{d\mathbf{L_{ij}}} = \mathbf{A_i}\mathbf{L_j}(t) \tag{5}$$

where $\mathbf{A_i}$ is the ith row of $\mathbf{A}$ and $\mathbf{L_j}(t)$ is the jth column of $\mathbf{L}(t)$. If vertex $\mathbf{v_i}$ is not a member of clique $\mathbf{c_j}$ then $r_{ij}(t+1) = 0$. The update equations for the label coefficients can now be defined as:

$$L_{ij}(t+1) = \frac{r_{ij}(t+1)}{\sum_k r_{ik}(t+1)} \tag{6}$$

This form of update can be viewed as a type of gradient accent since the value of $r_{ij}(t+1)$ is a function of the improvement of the score function associated with increasing $\mathbf{L_{ij}}$. Equation (6) ensures that the sum of the values in each row of $\mathbf{L}$ is equal to one.

Initially all label values for each vertex are uniformly distributed amongst their available cliques. After each iteration the value of β is gradually increased, in this way the label for the dominant clique for each vertex gets closer and closer to one and the rest of the labels approach zero. The estimate of the optimal label matrix can then be defined as: $\hat{\mathbf{L}} = \lim_{\beta \to \infty} \mathbf{L}_\beta$. Due to limited space we don't discuss the convergence properties of this method, interested readers are refereed to [3] for more details on this topic. A summary of the assignment algorithm is:

- for all i, j if $\mathbf{v_i} \subset \mathbf{c_j}$ set $\mathbf{L_{ij}(0)} = 1$ else set $\mathbf{L_{ij}(0)} = 0$
- for all i, j if $\mathbf{v_i} \subset \mathbf{c_j}$ set $\mathbf{L_{ij}(1)} = \frac{1.0}{\sum_k \mathbf{L_{ik}(0)}}$ else set $\mathbf{L_{ij}(1)} = 0$
- set $\beta = 1$
- set $t = 1$
- While not converged {
  - for all i, j set $r_{ij}(t+1) = e^{\beta \frac{dS(\mathbf{L}(t))}{d\mathbf{L_{ij}}}}$
  - for all i, j set $L_{ij}(t+1) = \frac{r_{ij}(t+1)}{\sum_k r_{ik}(t+1)}$
  - increase β
  - increment t by 1
- }

The bulk of the calculation is the $O(|V|)$ operations that must be performed to calculate r_{ij} at each iteration. There are $|C||V|$ elements in **L**. The size of **C** is bounded by $|\mathbf{V}|^2$ so the algorithm can be as complex as order $O(|V|^4)$ One interpretation of this

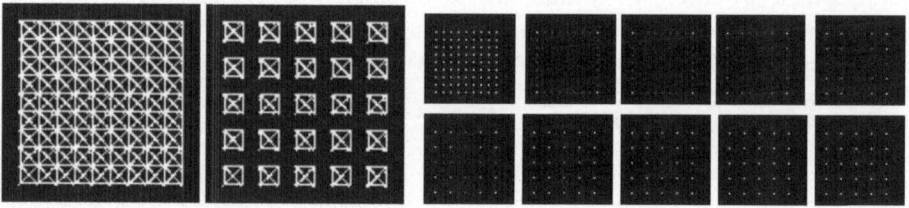

Fig. 1. Synthetic Experiment. Left shows the initial and final graph structure. The right starts by displaying all clique centers and the remaining images show the evolving clique strengths over time. Initially all the cliques are equally strong. However after the first iteration, the corner cliques quickly emerge and this starts a wave of certainty that propagates across the graph. This behavior is observed because vertices at the four corners of the graph can be viewed as seeds of certainty since they only have one maximal clique from which to choose

implementation of soft assign is that it is a form of relaxation labeling that propagates certainty across the graph. If a vertex is a member of a large number of maximal cliques, then based on local context there is a lot of ambiguity. This is often the case for vertices that are in the center of the foreground pixel cluster. However vertices near the periphery of the cluster may be associated with a relatively small number of cliques. These low ambiguity vertices help to strengthen their chosen cliques. As these cliques get stronger, they begin to dominate and attract the remaining less certain vertices. This weakens neighboring cliques which in turn lowers the ambiguity of vertices in that region. In effect this process is like peeling off the outer layer of the foreground region making it easier to segment the next outermost layer. This continues until clear segmentation penetrates right to the core.

To illustrate this behavior, a 10 by 10 grid of vertices is constructed. An allowable clique is limited to a width and height of 1 grid spacing so that edges only exist between vertically, horizontally and diagonally adjacent vertices. The emergence of the accepted cliques is illustrated in figure 1. Clearly this is the optimal solution since every vertex has been associated with a clique of maximum size.

3   Geometric Appearance Model

In section 2, it is argued that a geometric appearance model enables efficient enumeration of the clique structure of the graph **G**. A top view camera directly addresses issues associated with occlusion, however if a reasonably large area of coverage is desired and the site does not offer an extremely high vantage point, some perspective effects will be encountered. By taking advantage of the fact that most sites have a flat ground plane and that individuals are either standing or walking, a geometric bounding box can be generated as a function of the image coordinates.

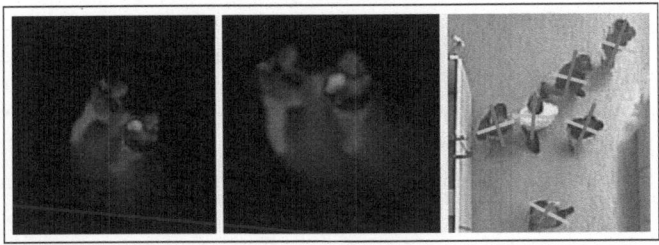

Fig. 2. Rectification of Foreground Patches. Left: Absolute background difference image in the partition. Middle: the image transformed in to rectified coordinates. Right: the height and width vectors across the image

Various authors [12] have taken advantage of the fact that people are roughly the same height and that they stand perpendicular to the ground. If this is the case, the foot plane and the head plane can then be defined. There exists two homographies H_f and H_h that maps the imaging plane to the foot and head plane. If the foot and head pixels p_f and p_h are from the same person then:

$$H_h p_h \propto H_f p_f \tag{7}$$

since the person is assumed to be standing perpendicular to the floor. A mapping between the foot pixel and the head pixel can then be defined as:

$$p_h \propto H_h^{-1} H_f p_f. \tag{8}$$

This mapping $H_h^{-1} H_f$ is also a homography.

When considering a foreground pixel patch, the center pixel is set to a foot pixel, and the head pixel is then determined via the homography $H_h^{-1} H_f$. The height vector goes from the foot pixel to the head pixel. We assume that the width of a person remains relatively constant across the image. The width vector is set to be perpendicular to the height vector. See figure 2. The local image is warped so that a person in this rectified view will be contained in a w by h bounding box. In general the head to foot mapping can be determined given a minimum of four head to foot pixel pairs [4].

The set of maximal cliques $\mathbf{C}$ can now be computed efficiently in this rectified coordinate system. Conceptually a window of size w by h can be slid across the foreground patch. At any time during this process, the vertices that are inside this window constitutes a clique. When ever there is a change in the set of interior vertices a new clique is formed. If this clique is maximal then it is added to $\mathbf{C}$. Unlike most graph problems, where one attempts to infer $\mathbf{C}$ given a graph with an existing edge structure, our process starts off with a set of vertices, $\mathbf{C}$ is then constructed and the edge structure of $\mathbf{G}$ can be inferred from $\mathbf{C}$.

4   Measurements

Given a partitioning function Ω which in this case takes the form of a grid of 15 by 15 pixel squares, a vertex for each partition can be defined by

$$\mathbf{v_i} = \max_{\mathbf{v} \in \Omega_i} [\mathbf{r} = |\nabla(|\mathbf{I} - \mathbf{B}| * \varphi_\delta)(\mathbf{v})|] \; , \tag{9}$$

where φ_δ is a suitable bandpass filter, I is the current image and B is the background image. Vertices with a value of r below a given threshold are rejected. An orientation vector is associated with each vertex. This is computed directly from the gradient of the absolute difference image. In general each person is surrounded by the background and it is assumed that most vertices will be located on the boundary of an individual. Since the absolute difference is computed, these boundary vertices should be pointing towards the center of the individual.

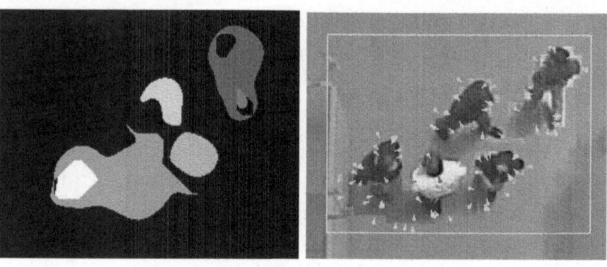

Fig. 3. Measurement Process. The image on the left shows the location and orientation of the interest points for a given foreground patch. Vertices for which the confidence in the orientation estimate is very low are marked in red. The image of the right show the label map which is used to estimate the confidence

In order to access the edge strength between two vertices, we hypothesize that both vertices are on the periphery of a person's outline. We also model the shape of a person as being roughly circular. Since the orientation of each vector should be pointing towards the center of the individual, the following model is defined:

$$\omega_j = \pi - \omega_i + 2\omega_{ij} \tag{10}$$

where ω_i is the orientation of vertex i, ω_j is the orientation of vertex j and ω_{ij} is the orientation of the line going from vertex i to vertex j. The strength $\mathbf{a_{ij}}$ of the edge is $\mathbf{e_{ij}}$ is defined as

$$\mathbf{a_{ij} = 1.0 - \frac{|\omega_j - (\pi - \omega_i + 2\omega_{ij})|}{\pi}} \tag{11}$$

5 Experiments

Figure 4 shows an example of the operation of the emergent labeling scheme. Initially a foreground pixel patch is identified. A rectified image is generated using the foot to head transform (see section 3). The gradient of the absolute background difference image is calculated and the oriented vertices are extracted. The set of maximal cliques C on G are then computed using the algorithm described in section 3. Edge weights are computed based on relative vertex position and orientation. The emergent labeling algorithm is executed resulting in an assignment of vertices to a subset of the cliques in C. The centroids of the significant cliques are transformed back to image coordinates

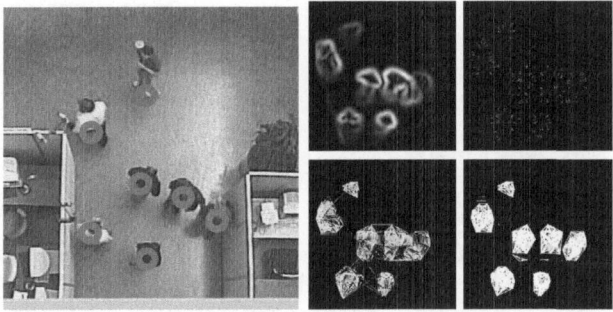

Fig. 4. First Example. The resulting state of the emergent labelling algorithm is shown on the left. The block of images on the right illustrates the most important intermediate steps. The gradient image, i.e. $|\nabla(I - B) * \varphi_\delta|$ is shown top middle. The orientation of the set of vertices are show on the upper right. The edge strength for the initial graph are displayed in the bottom middle. The final graph is show on the bottom right

and superimposed on the original imagery. In this example significance implies that at least half of the vertices that are members of a clique have been assigned to that clique. One of the challenging elements of this example is that the right hand pair of people are relatively close to one another. Upon inspection of the graph, the inter edge strengths between the vertices of these two individuals is quite strong making it difficult for standard clustering algorithms to function properly. The second example shown in figure 5 shows an even more challenging problem with four people positioned next to each other. A successful segmentation of an extremely crowded case is shown in the third example (see figure 5). An accompanying video shows the intermediary states taken to achieve this result. The main outlier appears to be a result of non human motion. A video sequence with over one hundred frames with seven moving individuals was segmented on a frame by frame basis (see accompanying video). It was found that on four occasions an individual was not detected.

In the next section, generating a graph from the image is viewed as a random process. This analysis will provide two things. First, we will develop a sense of how robust the algorithm is with respect to minor changes in the graph structure. This will lead to the identification of various failure modes. Second by sampling multiple graphs, an estimate of the prior $P(\mathbf{X})$ will be generated.

6   Generating $P(X)$

Since the partitioning function $\mathbf{L}$ and the associated state $\mathbf{X}$ are computed deterministically, it is the uncertainty of which interest points are associated with foreground objects and their orientation which needs to be captured. As it can be observed in figure 3 a number of interest points are caused by shadow regions. In addition the orientation associated with each vertex can be misleading. The aim of this section is to derive an acceptance probability that a vertex $\mathbf{v_i}$, given the magnitude of its response r (see equa-

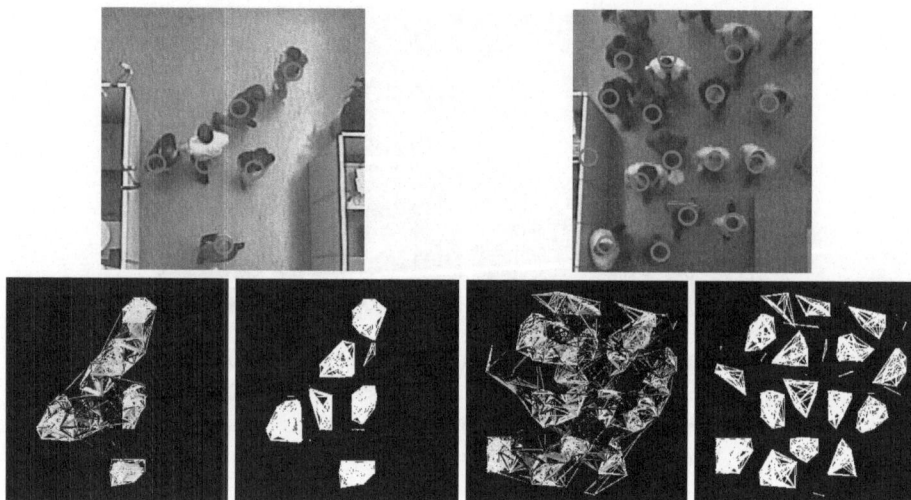

Fig. 5. Result. Two results of the emergent labelling algorithm are shown. The top row shows the estimated state for each of the experiments. Below each of these the initial edge strength are shown in the left. The estimated graph structure $\hat{\mathbf{L}}$, i.e. the result of the emergent labelling process is displayed on the right. The resulting clique centers or the estimated states $\mathbf{X}$ are shown on the right. Note that the algorithm generated one false detection in the example shown in the right column. Please refer to section 6 for further discussion

tion (9)), is a foreground vertex. Furthermore a confidence measure will be computed for the vertex orientation.

The purpose of computing $P(\mathbf{X})$ are twofold. Firstly $P(\mathbf{X})$ can effectively be used as an initialization prior for a tracking algorithm. Secondly, and perhaps more relevant to this work, this effectively allows us to study the robustness of the algorithm.

The acceptance probability for a vertex $\mathbf{v}$ given the feature response r can be written as

$$p(\mathbf{v} \in F | r) = \frac{p(r | \mathbf{v} \in F)\, p(F)}{p(r)} \qquad (12)$$

where F denotes the foreground area. The distributions $p(r | \mathbf{v} \in F)$, $p(F)$ and $p(r)$ are estimated from training data. The orientation, as defined in equation (10), is treated in a similar fashion. The confidence of the orientation estimation is based on the background/foreground separation of the pixel sites. The confidence is based on the minimal distance to a background pixel location. This model is sufficient for our case since we mainly suffer from confusions inside the foreground region.

In various experiments a number samples were computed by generating a vertex set V by sampling from the acceptance probability vertex as well as the acceptance probability of the vertex orientation. Whereas rejected interest points are excluded from the vertex set, the orientation of a vertex is flipped in case it is not accepted. The result of one particular case is summarized in figure 6. Most of the incorrect estimates are caused by wrongly associating interest points in shadow areas with foreground objects.

This particular experiment illustrates that the emergent labeling algorithm converges to a correct estimate in 65% of all cases. Based on our experiments for a number of different frames the sample population only rarely splits up into more than one mode. The additionally supplied videos support this claim. The experiment documented here also illustrates the fact the resulting prior $P(\mathbf{X})$ can effectively be used for tracker initialization.

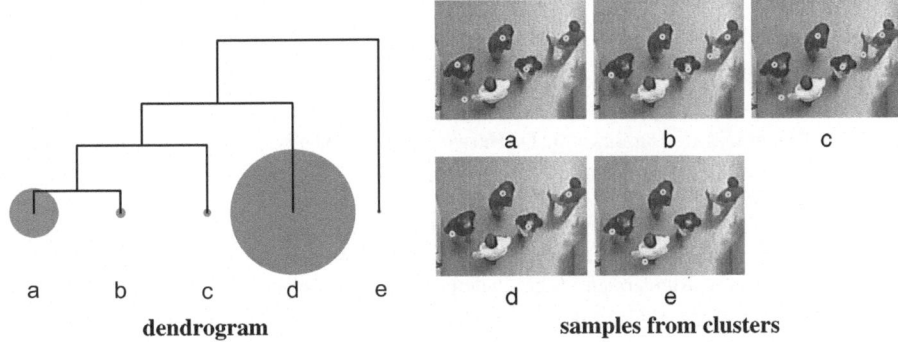

Fig. 6. Sampling Experiment. In an experiment 200 graph samples were generated for one frame in a test sequence. As described in the text each time a sample is generated the graph is computed from a different set of vertices and orientations. On the left the dendrogram of the entire sample population is shown. The size of each cluster is illustrated by the size of the red circles. Selected samples out of each of the clusters are shown on the right. The cluster d, which is by far the largest, only consists of correct estimates. Clusters a, b and c are caused by wrongly selected interest points cause by the shadow regions. The cluster e only contains one data point. In this case the algorithm does not detect one person at all

7   Discussion

The emergent labeling scheme is a new algorithm that can segment top view images of crowds. A graph theoretic approach was taken and by using a geometric shape model, the global score maximization problem is converted into a labeling exorcize which is solved using a version of the soft assign algorithm. An interpretation of this process is that it is a form of relaxation labeling which effectively propagates constraints from regions of low ambiguity to regions of high ambiguity. No prior information such as the number of individuals in the site is required and the algorithm runs in polynomial time.

It was observed that due to the dynamic nature of the problem, the graph generation mechanism should be viewed as a stochastic process and that a sampling paradigm can be used so as to generate a set of likely estimates of the state. This can be viewed as a form of importance sampling.

This segmentation paradigm can be extended in various directions. based on local image statistics each vertex could be described in a probabilistic terms as being some type of body part such as a head, a limb or a shoulder. The relationship between vertices would then have to be consistent with these descriptions. In this way the approach takes on a form of graphical model optimization.

Currently the system does not use any prior information such as previous segmentations and estimated appearance models of individuals in the scene. This type of information can be incorporated in to the cost function in terms of a prior on the state. It can also be used to generate more accurate estimates of edge strengths. In addition, the graphs themselves need not be restricted to vertices from the same image. Inter temporal edges could be constructed which would effectively fuse the tracking and detection problem.

References

[1] A. Elgammal, R. Duraiswami, D. Harwood, and L. Davis. Background and foreground modeling using nonparametric kernel density estimation for visual surveillance. *Proceedings of the IEEE*, 90(7):1151–1163, July 2002.

[2] D. Gavrila. Pedestrian detection from a moving vehicle. In *Proc. 6th European Conf. Computer Vision, Dublin, Ireland*, pages 37–49, 2000.

[3] S. Gold and A. Rangarajan. A graduated assignment algorithm for graph matching. *IEEE Trans. on Pattern Analysis and Machine Intelligence*, 18(4):377–388, April 1996.

[4] R. Hartley and A. Zisserman. *Multiple view geometry in computer vision*. Cambridge University Press, 2000.

[5] S.S. Intille, J.W. Davis, and A.F. Bobick. Real time closed world tracking. In *Proc 11th IEEE Computer Vision and Pattern Recognition, San Jaun, PR*, pages 697–703, 1997.

[6] M. Pelillo M. Pavan. A new graph-theoretric approach to clustering and segmentation. In *IEEE Computer Vision and Pattern Recognition, Madison, Wisconsin*, pages 145–152, 2003.

[7] S. Zucker M. Pelillo, K.Siddiqi. Matching hierarchical structures using association graphs. *IEEE Trans. on Pattern Analysis and Machine Intelligence*, 21(11):1105–1119, November 1999.

[8] A. Mittal and L.S. Davis. M2tracker: A multi-view approach to segmenting and tracking people in a cluttered scene using region-based stereo. In *Proc. 7th European Conf. Computer Vision, Kopenhagen, Danmark*, volume X, pages 18–33, 2002.

[9] C. Nakajima, M. Pontil, B. Heisele, and T. Poggio. People recognition in image sequences by supervised learning. In *MIT AI Memo*, 2000.

[10] R. Ronfard, C. Schmid, and B. Triggs. Learning to parse pictures of people. In *Proc. 7th European Conf. Computer Vision, Kopenhagen, Danmark*, volume 4, pages 700–710, 2002.

[11] J. Shi and J. Malik. Normalized cuts and image segmentation. *IEEE Trans. on Pattern Analysis and Machine Intelligence*, 22(8):888–905, August 2000.

[12] T. Zhao and R. Nevatia. Stochastic human segmentation from a static camera. In *IEEE Workshop on Motion and Video Computing, Orlando, FL, USA*, pages 9–14, 2002.

Author Index